T0074287

Human-Centered Social Media Analytics

Yun Fu

Editor

Human-Centered Social Media Analytics

 Springer

Editor
Yun Fu
Department of ECE, College of Engineering
Northeastern University
Boston, MA
USA

ISBN 978-3-319-05490-2   ISBN 978-3-319-05491-9   (eBook)
DOI 10.1007/978-3-319-05491-9
Springer Cham Heidelberg New York Dordrecht London

Library of Congress Control Number: 2014933527

Printed on acid-free paper

Springer is part of Springer Science+Business Media (www.springer.com)

Preface

With the advent of the social media era, technologies for social computing have become prevalent. This book provides a unique view of applying human-centered computing in the social media scenario, especially classification and recognition of attributes, demographics, contexts, and correlations of human information among unconstrained visual data from the social media domain. These cues are utilized for inferring people's social status, relationships, preferences, intentions, personalities, needs, networking and lifestyles, etc. Understanding of humans in social media will play an important role in many real-world applications with both academic and industrial values and broad impacts.

As a professional textbook and research monograph, this book comprises 10 chapters covering multiple emerging topics in this new interdisciplinary field, which links popular research fields of Face Recognition, Human-Centered Computing, Social Media, Image Processing, Pattern Recognition, Computer Vision, Big Data, and Human–Computer Interaction. Contributed by experts and practitioners from both academia and industry, these chapters complement each other from different angles and compose a solid overview of the human-centered social media analytics. Well-balanced contents for both theoretical analysis and real-world applications will benefit readers of different levels of expertise, help them to quickly gain the knowledge of fundamentals, and further inspire them toward insightful understanding. This book may be used as an excellent reference for researchers or as a major textbook for graduate student courses requiring minimal undergraduate prerequisites at academic institutions.

The content is divided into two parts of topics. The first five chapters are on Social Relationships in Human-Centered Media, while the last five chapters are on Human Attributes in Social Media Analytics. Chapter 1 provides an introduction to social relationship in the social media context, and describes how to bridge human-centered social media content across web domains; Chap. 2 presents a method for social relationships in media analytics, in particular, features, models, and analytics for learning social relations from videos; Chap. 3 discusses understanding social relationships in social networks. A method of community understanding in location-based social networks is presented; Chap. 4 describes the social relationship in social roles, especially social role recognition for human event understanding; Chap. 5 presents how to classify social relationships in human–object interactions through integrating randomization and discrimination;

Chap. 6 introduces a method to construct people recognition in social media through social context; Chap. 7 presents an example of demographic sensing in social media for female facial beauty attribute recognition and editing; Chap. 8 demonstrates face age estimation in social media from a data representation perspective; Chap. 9 presents methodologies of kin relationship and identity recognition and understanding from group photos in social media; Chap. 10 presents the application of a probabilistic model under social context to occupation recognition and profession prediction.

I thank all the chapter authors of this book for contributing their most recent research works. I also sincerely thank Simon Rees from Springer for support to this book project.

Boston, MA   Yun Fu

Contents

Part I
Social Relationships in Human-Centered Media

Chapter 1
Bridging Human-Centered Social Media Content Across Web Domains

Suman Deb Roy, Tao Mei and Wenjun Zeng

1.1 Introduction

Social media is at the forefront of all technologies that have had a disruptive impact on existing infrastructures. Being predominantly human-centered, social media has provided an innovative paradigm to address various cultural (e.g., freedom of speech), sociological (e.g., community opinions), and technological problems (e.g., media recommendation and popularity prediction) which were otherwise hard to address purely by traditional approaches. Real-time social data is being utilized for a variety of multimedia challenges, such as semantic video indexing, image/video context annotation, visualizing political activity, and tracking flu outbreaks [1]. Websites such as YouTube, Twitter, Facebook, Instagram, Reddit, Digg, and others flourish by engaging users with information leveraged from social media. Social streams like Twitter are an especially good indicator of crowdsourcing activity on the Web. The information in social streams is real-time, thus it can be used to learn about physical world events quickly [2]. Major world events in recent times, such as the Egyptian Revolution, the London riots, the Japan Earthquake, and the Boston bombings have been extensively captured using social media [3, 4].

However, media content on the Internet is unevenly distributed, often depending on platforms, popularity, and bias of web domains. The power of the media is thus confined by the domain where it originates. For example, video popularity is usually judged by view count alone, but not by how trending the video topic is—although

S. Deb Roy (✉)
Betaworks, 416 W 13th St, 203, New York, NY 10014, USA
e-mail: suman@betaworks.com

T. Mei
Microsoft Research, No. 5 Dan Ling Street, Beijing 100080, China
e-mail: tmei@microsoft.com

W. Zeng
University of Missouri, 119 Engineering Building West, Columbia, MO 65211, US
e-mail: zengw@missouri.edu

Y. Fu (ed.), *Human-Centered Social Media Analytics*,
DOI: 10.1007/978-3-319-05491-9_1, © Springer International Publishing Switzerland 2014

the latter seems to be an observable factor in relevancy. There are also various flavors of social media on the Internet—some publish valuable information, some share in real-time, and some provide crowdsourcing options. An important realization is that although social media has become a primal feature of the Web 2.0 era, it is essentially distributed disparately—spread across different web domains.

The connected power where each kind of media can enhance others has not been fully realized. A lot of interesting questions can be asked and answered. Do trends detected in social streams have latent relations with user search patterns in video publishing sites? If such intelligent associations can be drawn and analyzed, user experience in one media domain (e.g., social stream) can be enriched by utilizing information in another media domain (e.g., video publishing). This can help to solve some problems that sole multimedia techniques cannot accomplish elegantly, such as better modeling of video popularity using socially trending topics/events. Thus, we feel the need for better cross-domain media recommendation systems as a key constituent to *social search*; our goal is to empower online media with real-time social information aggregation from various sources. We will focus strongly on videos and tweets as examples of cross-domain media content.

One aspect of social streams like Twitter is its short text format, which is fast, real-time, and allows events to be instantly reported and broadcasted online. Consider this in the light of a traditional media application like video recommendation, e.g., a user is viewing an old video on "Egypt" (uploaded in 2008) on the eve of the "Egyptian revolutions" (Jan 29th 2011). Also, consider a journalist who just uploaded a live video of the revolution in YouTube. Most existing video recommendation systems have no way to *relate* these two videos, i.e., the seed video the user is watching and the related newly uploaded video, in spite of their striking similarity in topics. Both academic research and industrial software has proven that social data can significantly boost recommender systems. Social recommendation using social stream data has the potential to model factual world events in real-time by topic extraction and subsequently perform interesting tasks such as video associations among the old and fresh videos belonging to similar topics. Learning fresh video associations is important to improve the performance of multimedia applications, such as video recommendation in terms of topical relevance and popularity and timely suggestions.

As shown in Fig. 1.1, fresh and socially relevant (trending) videos (e.g. Egyptian Revolution-2012) can be associated to older videos (e.g., video documentary on Egyptian pyramids) to facilitate social trend aware recommendations. Social trend aware recommendation includes videos belonging to related trending topics for recommendation (a video about Egypt relates to breaking news event in Egypt).

This chapter focuses on two important techniques and three novel applications that utilize the potential of social media by algorithmic detection of trending topics, and utilizing the knowledge learned in developing socially aware multimedia applications. Realizing that there exists cross-correlation between media data in different domains often generated in response to the same events in the physical world, we first aim to build a common topic space between the domains of social stream and online video. In particular, we take Twitter as the exemplary social stream and the videos collected from a commercial video search engine as online video. The principal rea-

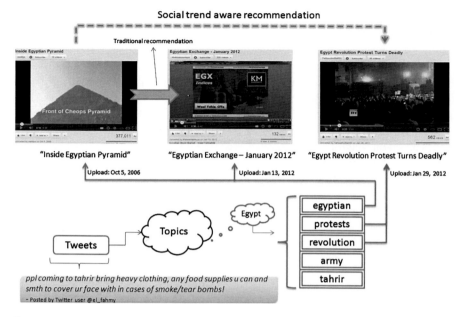

Fig. 1.1 Examples of learning topics from tweets that facilitate social trend aware video recommendation

son behind building a topic space is to construct a base context platform upon which various media applications can be forged. It acts like a *bidirectional* bridge between tweets and videos. We tweak the Latent Dirichlet Allocation (LDA) model to learn topics in real-time from the social stream [5]. The proposed topic model, called Online Streaming LDA (OSLDA), is utilized to extract, learn, populate, update, and curate the topic space in real-time, scaling with streaming tweets. Second, we discuss *SocialTransfer*—a scalable technique for real-time transfer learning between the domains of social streams and traditional media (e.g., video). *SocialTransfer* utilizes topics extracted from social streams to build an intermediate topic space in between the social and video domains. The topic space is an abstract space containing several clusters of words belonging to various topics that reflect world events in real-time (including current and past trends).

Based on these two techniques, three applications are discussed. First, a way to perform socialized video query suggestion, wherein topics from social media are used to suggest suitable search keywords based on user-provided search queries [6]. Second, a socially aware video recommendation framework that can find relevant videos based on what the user is currently viewing in some video publishing website [7]. The recommended video belongs to the topics that are being presently discussed in social media. Finally, we present a way to explain the unusual popularity bursts that some online videos enjoy throughout their lifetime [8].

1.2 Challenges

Exploring complex data from disparate resources and making sense of it is an
extremely challenging problem, especially when one of the resources is Twitter.
We need intelligent extraction of relevant and valuable information from social
streams (e.g., Twitter) and correlating social media across different domains. This
is a nontrivial problem, primarily due to the noisy nature of social streams. Moreover, each tweet in Twitter is limited to 140 characters. This severely hinders techniques based on "bag-of-words." Tweets are often noisy and improperly structured in
grammar/syntax, which makes them difficult to process using standard Natural Language Processing (NLP) tools. Social stream data typically arrives in high-volume
(bursty traffic) and thus, algorithms mining them must scale in learning. For example,
some methods based on Singular Value Decomposition (SVD) are too slow to scale
[9].

We attempt to overcome other challenges for intelligent and productive usage of
social media, and to build a bridged connection among the disparate social media on
the Internet, using which cross-domain media recommendation can be realized. The
main concern in this regard is to transfer knowledge across domains, which requires
aligning features common to both domains (e.g., video tags and social stream topic
words as shown in Fig. 1.1). Specifically, there are four distinct challenges in cross-
domain socialized recommendations:

1. A unified framework to combine the social and multimedia feature information
   which has different domain-specific properties.
2. A transfer learning algorithm that can seamlessly propagate the knowledge (i.e.,
   social topics) mined from the crowdsourced social streams to the video domain.
3. The scaling up and adaptation of the transfer learning algorithm to the ever bursty
   real-time nature of the social streams.
4. Dealing with the noisy, incomplete, ambiguous, and short form nature of social
   stream data. Each tweet is limited to 140 characters and often improperly structured in grammar/syntax. Traditional language model (e.g., Bag-of-Words) would
   fail to scale up with such kind of data.

In order to find the social popularity of a video (or image), we need to address
some other issues. In our model, social popularity depends on the *social prominence
of media content* in various social media domains on the Internet based on the visibility of the media topic in online social networks. We also need to gather empirical
evidence that media in some social media domains gains bursty/sudden popularity
due to the increased popularity of the media topic in another online social network
domain which would prove that the popularity signal is carried across domains of
social media existence (Twitter to YouTube). One of the applications presented here is
showing that the *SocialTransfer* algorithm is capable of carrying those signals across
domains of social media. Our large-scale experiments on social and video data are
especially catered to test the effectiveness of the cross-domain popularity penetration
hypothesis, i.e., how social network topic popularity affects the popularity of media
with the same topic across disparate social media domains.

1.3 Learning Topics from Social Streams

Our goal is to build a common topic space between the domain of social stream (*tweetSide*) and video domain (*vidSide*), where tweets are connected to topics, and topics to videos. Figure 1.2 shows how we envision this. Tweets are preprocessed, and then topics are extracted to fill up the topic space from *tweetSide*. On the *vidSide*, we have clicked through video data from a commercial search engine. First, we must extract video attributes/features (such as video tags, words in description and title, etc.). Then we create a record level inverted index. This index is further divided by categories (based on the 16 YouTube categories) for easier indexing.

We can use online learning Latent Dirichlet Allocation (LDA) to extract topics ($z \in Z$) from a stream of tweets ($d \in D$). LDA generates two distributions: a topical word-topic distribution $P(w|Z)$ and topics-tweets distribution $P(z|d)$. The vocabulary consists of words $w \in W$. Parameters α and are β Dirichlet priors to the topic-tweet and the word-topic distributions respectively. A tweet is a sequence of words, where w_n is the nth word in the sequence. Consider a k-dimensional Dirichlet random variable that can takes values in $(k - 1)$ simplex. LDA assumes the following generative process for each tweet d in the corpus D: (i) Choose N from a Poisson distribution. (ii) Choose $\theta \sim Dirichlet(\alpha)$ (iii) For each of the N words w_n: (a) Choose a topic $z_n \sim Multinomial(\theta)$ (b) Choose a word w_n from $p(w_n|z_n, \beta)$, a multinomial probability conditioned on the topic z_n. The dimensionality k of the Dirichlet distribution is assumed known and fixed.

Therefore, the joint distribution of the topic mixture θ, the set of N topics Z, and a set of N words in the vocabulary W is given by:

$$p(\theta, Z, W|\alpha, \beta) = p(\theta|\alpha) \prod_{n=1}^{N} p(z_n|\theta) p(w_n|z_n, \beta). \quad (1.1)$$

LDA represents every tweet as a random mixture over latent topics, whereas every topic has a distribution over the words. A topic comprises a set of *topical words*. For example, one topic generated by LDA is: {*egypt, mubarak, tahrir, army, revolution,* ...} , which clearly is related to the concept of the Egyptian revolution in Feb, 2011. On the *vidSide*, we have a set of videos (*V*) with related video identifiers. Our goal is to find the membership strength each video possesses with the set of topics in the topic space.

However, remember the two constraints of tweets—speed and variability. Our system learns in real-time by updating the topic space with every incoming stream of Tweets in a time slot. We call it Online Streaming LDA (OSLDA), since it leverages online LDA and also scales across streams of incoming tweets, updating tweet-topic and topic-video connections at the same time [6]. Unlike some previous variations of LDA, where updates are made on the word-topic prior distribution with β time, our method updates the topic space with time, using an active time decay function. Thus, OSLDA assumes the word-topic distribution can change significantly due to the dynamic nature of tweets. This makes our model robust.

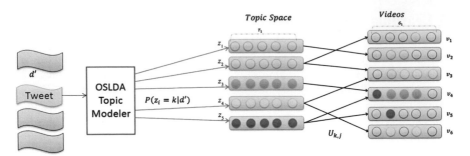

Fig. 1.2 OSLDA learning topics, populating topic space, and connecting to videos

With each time slot, OSLDA models incoming bursts of tweets and updates the topic space. Empirical studies showed that fixing number of topics to 30 was enough for 60 K tweets per time slot. Intuitively, processing more tweets should take more time, but the number of topics needed to be extracted from a sudden burst (say 120 K tweets) is usually less, since the burst is typically caused by a single event (single topic). So, the number of topics to be extracted does not double if the tweet burst doubles. This observation of social stream information dynamics allows us to focus what we want to extract and scale the algorithm to bursts of tweet data.

Thus, we can envision a *bidirectional connection* between tweet and video domain using the topic space, which is used to enrich tweets with video recommendations and enrich videos with social popularity. We use the Online Streaming LDA (OSLDA) model for real-time topic learning from Twitter stream as shown in Fig. 1.2. Each topic comprises a group of related words called *topical* words. Topic learning treats each tweet as a document and builds a generative model to connect the tweet to one or more topics. Thus, the topic of a tweet contains words (topical words) that are related to the tweet words but might not be explicitly present in the tweet itself. More precisely, the topic modeling generates two distributions, a tweet-topic distribution and a topic-word distribution.

Note that a video tag is a video identifier. For the jth video, the set of tags is represented by G_j. We also have a set of topical words (which were already extracted from tweets). Let the topical words in the kth topic be represented by the set T_k. Then, treating the set of topics and videos as a bipartite graph, we can define a link weighting function U such that:

$$U_{k,j} = \frac{T_k \cap G_j}{T_k}, \quad 0 \leq k < |Z|, \ 0 \leq j < |V|. \tag{1.2}$$

Thus, the more the common tags a video has with the words of a topic, the higher the weight $U_{k,j}$; and thus the higher the membership of the video toward this topic.

1.4 The *SocialTransfer* Framework

SocialTransfer is a unique method to combine topic modeling and transfer learning; providing a natural interface for topic modeling to seamlessly fit into the process of transfer learning. It uses OSLDA to populate a graph called the transfer graph, which itself is capable of updating itself with new tweets stream topics in real-time. The transfer graph's main purpose is to capture the cross-domain attributes of social streams and videos for using in the transfer learning task and model the relation between the auxiliary data from Twitter and the target video data. This "transfer graph" (Fig. 1.3) contains the instances, features, and class labels of the target data and the observed auxiliary data as vertices. The edges are set up based on the relations between the auxiliary and the target data nodes. The transfer graph presents a unified graph structure to represent the task of transfer learning from social domain to video domain.

Since every topical word in the topic space has an assigned topic label, the entire topic space can be treated as some sort of social bias for any semi-supervised learning task that requires social influence. This serves as a natural way to incorporate this social bias into transfer learning. Note that each assigned topic consists of a cluster of topical words. Similarly, each topic can be considered a cluster in the topic space. We limit ourselves to incorporating only selected topics from the topic space as input supervision (an additional set of labeled instances) for the transfer learning task. This choice will depend on factors such as whether we want to model only fresh (trending) topics or only video category-specific topics. Thus for K topics in the global topic space, we can choose a particular set of topical words $A_i^{in} \subseteq A_i$, for $i = 1, 2, \ldots, K$ to act as the bias or input supervision to update the transfer graph before spectral learning. This sort of input topic supervision is fed into the transfer graph progressively, as depicted in Fig. 1.4, where topics modeled in real-time from the social stream using OSLDA are used to update the transfer graph by means of a ranked update (Eq. 1.5) on the transfer Laplacian matrix representation of the transfer graph. This allows progressive and seamless inclusion of topics into the transfer graph as shown in Fig. 1.4, facilitating the social influence in transfer learning.

The novelty of our approach lies in how we incorporate the learned social topics into this transfer graph in real-time. This task is nontrivial, since if not properly done, it may incur substantial costs in terms of scalability (e.g., in eigen-feature extraction) and interoperability (in integration of topics) between topic modeling and transfer learning. As described later, we incorporate the learned topic model into the transfer graph by means of a ranked update on the Laplacian matrix representation of the transfer graph.

Let us focus on the example in the transfer graph illustrated in Fig. 1.3. The feature word "recyclopath"[1] occurs in the training video instance "Interview with Mel Kelly (aka Recyclopath)" shown in the top right. Since the video lacks any tags related to "Environment," a traditional learner will find it difficult to extract the topic of

---

[1] Recyclopath means a person who is almost paranoid about recycling and is an extreme environmentalist.

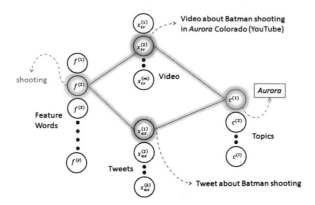

Fig. 1.3 The transfer graph consisting of vertices from auxiliary, training, and test data and edges indicating co-occurrence of word/topic in a tweet or video

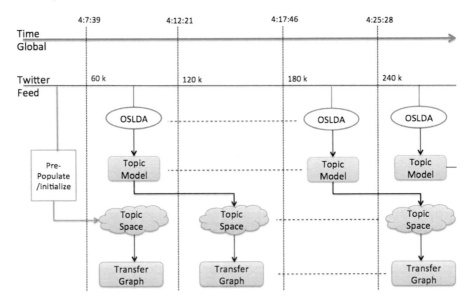

Fig. 1.4 Real-time topics being fed into the transfer graph

this video to be related to "Environment." However, the auxiliary data has a tweet instance belonging to the "Environment" topic having the word "recyclopath." Thus, the transfer learner can label this video as "Environment"-related and associate this video to another "Environment"-related video. This is an example of discovery of video associations by understanding video topics with the help of social topics.

As shown in Fig. 1.3, the transfer graph $G(V, E)$ consists of vertices representing instances, features or class labels, and edges E denoting co-occurrences between end nodes in the target and the auxiliary data, defined as follows, where χ_{aux} represents auxiliary Twitter data, χ_{train} and χ_{test} represents videos, $\mathcal{F}$ represents the feature word space (obtained from videos/tweets) while $\mathbb{C}$ is the category label of topics.

$$V = \chi_{\text{train}} \cup \chi_{\text{test}} \cup \chi_{\text{aux}} \cup \mathcal{F} \cup \mathbb{C}. \tag{1.3}$$

The weight of each edge where one of the end nodes belongs to $\mathbb{C}$ indicates the number of such co-occurrences. Let $\omega_{x,f}$ represent the importance of the feature $f \in \mathcal{F}$ that appears in instance $x \in \chi_{\text{train}} \cup \chi_{\text{test}} \cup \chi_{\text{aux}}$. Then, the weight of an edge where one of the end nodes belongs to $\mathcal{F}$ is indicated by $\omega_{x,f}$. The importance of a feature word $\omega_{x,f}$ can be calculated using the topic-word probability distribution matrix obtained from OSLDA. The total number of features and class label nodes remains fixed in the transfer graph. Let $T(x)$ represent the true label of the instance. If e_{ij} denotes the the weight of an edge between two nodes ϑ_i and ϑ_j in the transfer graph, then edge weights can be assigned as:

$$e_{ij} = \begin{cases} \omega_{\vartheta_i,\vartheta_j} & \vartheta_i \in \chi_{\text{train}} \cup \chi_{\text{test}} \cup \chi_{\text{aux}} \bigwedge \vartheta_j \in \mathcal{F} \\ \omega_{\vartheta_j,\vartheta_i} & \vartheta_i \in \mathcal{F} \bigwedge \vartheta_j \in \chi_{\text{train}} \cup \chi_{\text{test}} \cup \chi_{\text{aux}} \\ 1 & \vartheta_i \in \chi_{\text{train}} \bigwedge \vartheta_j \in \mathbb{C} \bigwedge T(\vartheta_i) = \vartheta_j \\ 1 & \vartheta_i \in \chi_{\text{aux}} \bigwedge \vartheta_j \in \mathbb{C} \bigwedge T(\vartheta_i) = \vartheta_j \\ 1 & \vartheta_i \in \mathbb{C} \bigwedge \vartheta_j \in \chi_{\text{train}} \bigwedge T(\vartheta_j) = \vartheta_i \\ 1 & \vartheta_i \in \mathbb{C} \bigwedge \vartheta_j \in \chi_{\text{aux}} \bigwedge T(\vartheta_j) = \vartheta_i. \end{cases} \tag{1.4}$$

For all other cases except the ones mentioned in Eq. (1.4), we set $e_{ij} = 0$. The edge weights thus represent the occurrence/importance of a category or feature present in the auxiliary/target data, which will be eventually utilized as a distance metric during spectral clustering. Some nodes in the graph may be isolated with no edge connections. The matrix updating process (discussed later) adds new edges to the isolated nodes. The transfer graph G is usually sparse, symmetric, real, and positive semi-definite, which allows the possibility of calculating its spectra efficiently [10]. The graph spectrum in terms of eigenvectors is the impression of the structure of relations among the source and target data. This structural relation between the cross domain data is the essence of transfer learning [11]. Thus, it is necessary to represent the source and target data as a transfer graph and then analyze their structural relation by learning the graph spectrum.

Once the transfer graph $G = (V, E)$ is built, we can use graph spectra analysis to form an eigen feature representation, which combines the principal component features from the training and the auxiliary data. In order to extract the top-q eigenvectors of the transfer graph $G = (V, E)$, we first need to convert the graph into a Laplacian matrix. Let $\deg(\vartheta_i)$ denote the degree of the i-th vertex in G. Then the transfer graph Laplacian $L_{\text{input}} := (l'_{i,j})_{|V| \times |V|}$, can be obtained as:

$$l'_{i,j} := \begin{cases} \deg(\vartheta_i) & \text{if } i = j \\ -1 & \text{if } i \neq j \bigwedge e_{ij} = 1 \\ 0 & \text{otherwise.} \end{cases} \tag{1.5}$$

If the Laplacian eigenvalues are represented as: $\lambda_0 = 1 \geq \lambda_1 \geq \cdots \geq \lambda_p$, then the eigen gap can be defined as: $eigengap = \frac{\lambda_q}{\lambda_q - 1}$.

Since the Twitter stream is extremely dynamic, topics and trends change over time. This requires a feature extraction scheme that can reflect and scale with the social stream. Previous approaches for spectral feature representation in transfer learning have suggested the use of the normalized cut (Ncut) technique for eigenvector extraction. However, our experiments showed that the normalized cut technique is incapable of scaling with the twitter stream [6]. Therefore, we use a Power Iteration technique for computing the q largest eigenvectors of L_{input} [12]. The method begins with a random $|V| \times q$ eigenvector matrix and iteratively performs matrix multiplication and orthonormalization until convergence. The speed of convergence of this method depends on the eigen gap, i.e., the difference between successive eigenvalues. In fact, it has been previously found that the number of steps required for the orthogonal convergence in the Power Iteration method is $O(\frac{1}{1-eigengap})$.

Since topics are updated in the topic space with time, we need to devise a way to progressively incorporate these new topics into the transfer graph. These topics could be incorporated by picturing them to be a time-dependent labeled bias (like a semi-supervised bias) which is an additional set of labeled instances acting as input supervision. One option for incorporating the semi-supervised topic bias as input supervision into the Laplacian representation of the transfer graph (L_{input}) is by producing a ranked update on L_{input} (see Eq. 1.7). The update in effect recalculates the weights of edge/path between the features and the corresponding labels within the transfer graph, thus updating the characteristic of the Laplacian (Eqs. 1.6, 1.7). Essentially, the ranked update on the Laplacian using the topic bias adds positive weights between feature words that share the same topic and adds negative weights between feature words that belong to different topics. Thus, the target and the auxiliary data instances act as a sort of virtual nodes enabling this reweighing of the feature edges.

An additional reason for using the ranked update technique has also rigorously demonstrated that when Laplacians such as L_{input} is positive semi-definite, a ranked update can improve eigenvector extraction speed by spreading the eigen gap [13]. The next subsection elaborates on how we use ranked updates to incorporate semi-supervised topic bias and update the transfer Laplacian.

We know from topic modeling that the words in tweets can be clustered into topics. Let us consider there are K such topic clusters. The semi-supervised topic bias is implemented by assuming we know the correct topic labels for a subset of the feature words. Said in terms of the transfer graph, the supervised bias is an additional input of the correct cluster labels for a subset of the feature vertices. This input is learned by topic modeling using OSLDA. The semi-supervised bias consists of a set of topical words for each topic $A_i^{in} \subseteq A_i$, for $i = 1, 2, \ldots, k$ that act as input supervision. Let us consider the simple case of two topic clusters A_1^{in} and A_2^{in}, such that $A^{in} = A_1^{in} \cup A_2^{in}$ denotes the set of labeled bias instances. Also, consider $d_i = \sum_j e_{ij}$ and $\text{vol}(A_k) = \sum_{i \in A_k} d_i$. We can then define a regularization vector δ_1 as:

$$\delta_1\,(i) = \begin{cases} \sqrt{\dfrac{d_i}{\mathrm{vol}(A^{\mathrm{in}})}} f\,(i)\,, & i \in A^{\mathrm{in}} \\ 0, & i \notin A^{\mathrm{in}} \end{cases} \qquad (1.6)$$

where, $f\,(i) = \sqrt{\dfrac{\mathrm{vol}(A_2^{\mathrm{in}})}{\mathrm{vol}(A_1^{\mathrm{in}})}}$ if $i \in A_1^{\mathrm{in}}$ and $f\,(i) = -\sqrt{\dfrac{\mathrm{vol}(A_1^{\mathrm{in}})}{\mathrm{vol}(A_2^{\mathrm{in}})}}$ if $i \in A_2^{\mathrm{in}}$.

The effect of the above Eq. (1.6) is to introduce a quadratic penalty if there is a violation in the topic bias label constraints. In other words, this will cause vertices of features that belong to the same topic to cluster together while vertices of different topics will be assigned to separate clusters (due to the penalty).

A rank-1 update on the original Laplacian can be made as:

$$L_{\mathrm{topic\_bias}} = L_{\mathrm{input}} + \gamma \cdot \delta_1 \delta_1^T. \qquad (1.7)$$

Similarly, if there are K topics, we can modify the original matrix L_{input} with a rank-k update [15] instead of a rank-1 update. This supervised ranked update first allows us to seamlessly incorporate streaming data progressively. Second, it aims at tuning certain algebraic properties of the input Laplacian matrix which are related to the convergence rate of the Power Iteration method, eventually speeding the eigen decomposition.

In summary, the input supervision using topics learned from the social stream allows us to implement rank-k updates on the transfer-Laplacian matrix as a similarity learning mechanism, where vertex similarities are adjusted on the basis of the topic bias. Note that the number of nodes in the graph is not changed during updating (dimension $|V|$ is fixed); instead the updates only introduce new edges or re-weights existing edges in the graph as it iteratively reuses the eigenvectors from previous update. Due to lack of space, we refrain from describing in detail how the rank-k update improves the speed of eigenvector extraction. Essentially, the ranked update increases the eigen gap, which accelerates the convergence of the Power Iteration method.

1.5 Predicting Bursty Video Popularity Using Social Media

The bursty or sudden rise in popularity of a video observed in the video domain can be largely attributed to the social importance of the video topic in the Twitter sphere, as shown in Fig. 1.5. This means before modeling popularity, we should first be able to classify a video as having a certain membership score to each topic in the intermediate topic space. We can utilize *SocialTransfer* in calculating the *social prominence* of a video and estimate its social popularity using the following two steps:

1. calculate the trending score for each topic (called Tscore) and use *SocialTransfer* classification to find the principal topic of a video. The trending score of the principal topic of a video is its *social prominence*;

Fig. 1.5 a Common
patterns of video popularity
on YouTube (based on view
count). **b** Twitter's effect in
detecting real-world events,
which translates to more
searches about the trending
topic leading to more video
hits

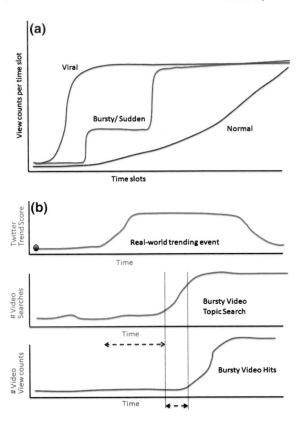

2. fusing *social prominence of a video* with its traditional popularity (based on view
   count) to estimate the final trend aware popularity score (*TAP*).

   Trends are temporal dynamic entities, meaning they grow for a certain period
of time, after which they suffer inevitable decay. In other words, trends remain
socially prominent for some time and their attractiveness fades away. It is therefore
necessary to include a time decay factor when modeling the trending score. To
describe mathematically, consider *SocialTransfer* receives a set of D tweets in one
time slot; t_{cur} being the current time slot and t_{onset} is the time slot when the trend
was first observed. We can then define the trending score of a topic z as:

$$\text{Tscore}_z = \frac{\sum_{k=1}^{|D|} P\left(z|d_k, t_{cur}\right)}{|D| \cdot \delta_z} \tag{1.8}$$

where $\delta_z = \varphi\left(t_{cur}, t_{onset}\right)$ is the time-dependent decay factor which is a function of
the current time slot and the time slot when the trend was first seen. The decay factor
must actively respond to trend reoccurrences (i.e. when the trend rises after an initial
fall). The decay can be formulated as:

$$\text{tr} = \begin{cases} 1, & P\,(z|D,\ t_{\text{cur}}) \geq P\,(z|D,\ t_{\text{cur}} - 1) \\ 0, & P\,(z|D,\ t_{\text{cur}}) < P\,(z|D,\ t_{\text{cur}} - 1) \end{cases}$$

$$\delta_z = \begin{cases} 1, & t_{\text{cur}} = t_{\text{onset}} \\ \delta_z, & t_{\text{cur}} > t_{\text{onset}} \text{ and } \text{tr} = 1 \\ \delta_z + \eta, & t_{\text{cur}} > t_{\text{onset}} \text{ and } \text{tr} = 0 \end{cases} \tag{1.9}$$

where $0 < \eta \leq 1$ depends on the category of the topic z (meme, music etc.). In addition to the usual trends, active decay can capture extremely dynamic trends like memes or sports related topics, which have short life spans compared to music or entertainment-related trends. Our final goal in this work is predicting which videos will demonstrate bursty nature based on their *TAP*.

For some video v, let z_v^* be the topic to which the video has maximum membership. This membership measure can be easily retrieved using SocialTransfer classification, since the output of the classification is the topic of the video. Then the *social prominence* of video v is $\text{Tscore}_{z_v^*}$:

In a traditional video ranking system (like in YouTube) videos with higher view counts are boosted in the rank list [14]. Thus, these videos get clicked more often, resulting in subsequent higher view counts for them. Therefore, it is necessary to engineer a reasonable fusion of the traditional approach and our proposed social prominence approach. This fusion of the traditional popularity factors (like view counts) and the social prominence of the video is called the Trend Aware Popularity (*TAP*).

In formulating the final popularity score, we also need to take into account the time when the video was uploaded (t_{upl}) since we need to discount the fact that older videos already have higher view counts. Thus, the net temporal Trend Aware Popularity score that we assign to a video v is:

$$TAP_v = \gamma \cdot \text{Tscore}_{z_v^*} + (1 - \gamma) \cdot \frac{t_{\text{onset}} - t_{\text{upl}}}{t_{\text{cur}} - t_{\text{upl}}} \cdot \#\,(\text{vc})_{t_{\text{onset}}} \tag{1.10}$$

where $\#\,(\text{vc})_t$ represents the view count at time t and γ is a weighting factor that balances social versus traditional popularity control. The above equation measures the social trend aware popularity of a video. The traditional popularity is reflected by the adjusted view count measure, which fractions the view count of a video based on when the video was uploaded in video domain, when the video topic trend was onset in social domain, and when the prediction was performed.

The *TAP* score reflects the social popularity as well as the traditional (video domain) popularity for a certain video. Our hypothesis is that social popularity signal penetrates across media domains on the Internet. In other words, if a topic is substantially popular (trending) in the social domain, then media belonging to the same topic will gain popularity in other domains (in this case, video domain). Therefore, a ratio of *TAP* to a scaled $\text{Tscore}_{z_v^*}$ value will provide us with the quantitative estimation of the impact of the social signal in boosting the overall video popularity for some video v. The lower the value of this ratio, the higher the impact of the social prominence of the video in comparison to the adjusted view count score. Given the same social prominence, the ratio seems to favor videos with lower adjusted view count measure. However, this is not an issue, since the adjusted view count measure is lower when the trend has been seen for longer time period ($t_{cur} - t_{onset}$), which practically means that we are more sure of the prediction if we are exposed to more of past trend data. Thus, for a certain video, if this ratio is significantly lower than for others (lower 10th percentile), we predict the video will gain bursty popularity.

1.6 Application Performance

Social Query Suggestion: The first set of experiments is conducted using video query logs from a commercial video search engine and 10.2 million tweet data. The goal was to find a temporal pattern or common terms between tweet topic words and video search keywords from video logs. From the data, we observed that there is a few minutes' time lag between a trend topic appearing on Twitter, and the same topical words being searched on the commercial video search engine. This means as trends rise and fall in Twitter, the volume of queries on the same topic rises and falls for video search, as shown in Fig. 1.6. To further support our claim that people search for Twitter trends outside Twitter, we also analyzed the query keywords used in a commercial video search engine on Feb 11, 2011. From Fig. 1.7, we can observe that if we eliminate daily searches such as "cats," "movies," "funny commercials" which are common, then topical words (detected by OSLDA) occupy a significant portion of the remaining video search keywords. In the video search engine logs and for all queries on Feb 11, which are not daily search terms (like 'cats'), 63 % of query words were detected by OSLDA.

In fact, this technique of socialized query suggestion can be extended beyond video search. We used Google Insights to understand search patterns for Web and image search on Feb 11, 2011. It was not surprising that "*Egypt*" was the hottest search topic that day. In fact, Google Web Insights provided us with the top 10 web search keywords related to "*Egypt*"; 7 of which had already been detected by OSLDA earlier. For Google Image search, 6 of the top 10 search keywords were detected by OSLDA. This is convincing evidence that the OSLDA detects relevant socially active topics within the *SocialTransfer* framework.

Socially aware Video prediction: We test *SocialTransfer* against a traditional learner like Support Vector Machines, where *SocialTransfer* uses auxiliary social data in combination with training data, whereas a traditional learner uses only the training data for prediction (called Non-Transfer) and serves as our benchmark. The target dataset consists of 5.7 million videos in total along with their RVGs (contains a list of related videos for each seed video). Our training data consists of 60 % videos randomly picked from the 4.8 million YouTube videos. The rest 40 % videos ($\sim$2 million) are used for testing. As auxiliary data, we use the 10.2 million tweets from the Twitter stream. We ensure to extract topics from tweets based on approximately 90 categories (16 main + 75 other) so that the source and target domains share the same categories. We observed and analyzed the average error in prediction for the

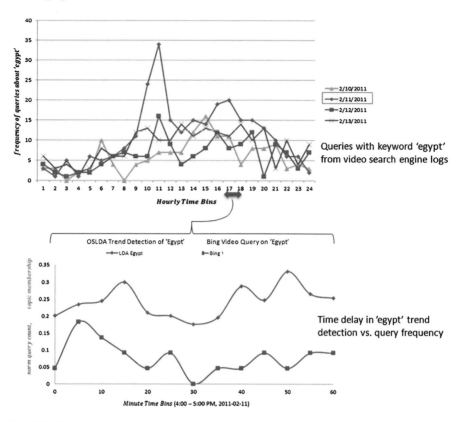

Fig. 1.6 The time delay between detection of Twitter trends using OSLDA and query frequency rise in video search engines can be used for query suggestion and pre-caching

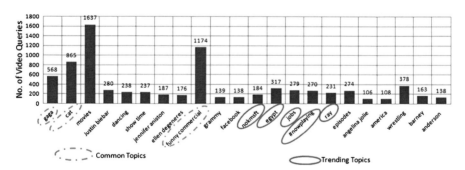

Fig. 1.7 This figure, taken from the [7], shows that a significant number of video search query keywords originate from topical words computed by OSLDA

Non-Transfer cases (SVM on training only) versus *SocialTransfer*. Non-Transfer refers to application of the traditional SVM learner to the original target dataset with no social influence (only training features are used); *SocialTransfer* means to apply SVM on the combined feature representation learned using transfer learning from social data (training + auxiliary). The performance is measured in error rate by averaging 10 random repeats on each dataset by the two evaluation methods. For each repeat, we randomly select 5,000 instances per category as target training data. We report the prediction error rate in each of the main categories, along with the overall error for the entire dataset. The overall gain using *SocialTransfer* is ∼35.1 % compared to non-transfer cases.

Social Video Popularity Prediction: In this case, the overall gain using *Social-Transfer* is ∼39.9 % compared to non-transfer cases. Performance improvement using transfer learning is most in category "Music." In all the major categories, *SocialTransfer* performs better than a traditional non-transfer learner. The F1-score of positive bursty videos in the proposed *SocialTransfer* algorithm is 0.68 whereas for the non-transfer SVM it was 0.32.

Additionally, we ran a baseline Naive Bayes classifier, which produces an F1 score of 0.21 without any transfer of auxiliary data. If we replace the SVM in *SocialTransfer* with the Naive Bayes, the F1 score achieved is 0.49. The drop in performance of Naive Bayes in both transfer and non-transfer cases compared to SVM (-0.19 and -0.11 respectively) is expected. Naive Bayes is easy to implement, but it suffers from strong feature independence assumptions. Notice that this feature independence assumption is more costly in the transfer scenario, where the drop in performance is larger than in non-transfer scenario, potentially due to the heavy reliance of *SocialTransfer* on cross-domain feature alignment.

1.7 Conclusion

Social data is rich in information, and yet, noisy in construction where the strong signal is difficult to find. Building novel algorithms that algorithmically analyze social data is pillared upon a fundamental understanding of information diffusion dynamics in social media. Such insight is often found through network analysis of the online social network topography, NLP on social textual data, and signal analysis of the spatiotemporal trends in social media.

In this chapter, we present two interesting techniques based on fundamental observations of the online social world. Our goal is to augment traditional machine learning techniques with intelligent socially aware algorithms. First, we realize that a burst of social media activity usually indicates a single topic, which reduces the overhead of computing several topics when performing topic modeling even if there are an increased number of documents. This technique is OSLDA. Second, we understand the richness of social data lies in its real-time nature. We place this as a high level priority in our design of *SocialTransfer*. SocialTransfer can update the transfer graph with newly learned topics, and it does so by elegantly scaling with the tweet bursts.

Finally, we also discuss three applications built on top of the two techniques. They are all socially aware, augmenting traditional applications with social media trends information. The first application allows us to predict socially relevant queries based on what the user is searching. The second allows the user to get related recommendation on videos that belong to socially trending topics. The final application deciphers why some videos on YouTube receive bursty popularity at those certain points in their lifetime, theorizing that at those points the video topic is of potential social interest. Social media has been often referred to as a signal for the human condition, and socially aware algorithms of the future will be able to measure the pulse of the web content by intelligently analyzing social multimedia data in real-time.

References

1. Kaplan, A.M., Haenlein, M.: Users of the world, unite! The challenges and opportunities of social media. Bus. Horiz. **53**(1), 59–68 (2010)
2. Roy, S.D., Mei, T., Zeng, W., Li, S.: Towards cross domain learning for social video popularity prediction. IEEE Trans. Multimedia **15**(6), 1255–1267 (2013)
3. Lotan, G., Graeff, E., Ananny, M., Gaffney, D., Pearce, I., Boyd, D.: The revolutions were tweeted: information flows during the 2011 Tunisian and Egyptian revolutions. Int. J. Commun. **5**, 1375–1405 (2011)
4. Tonkin, E., Pfeiffer, H.D., Tourte, G.: Twitter, information sharing and the London riots? Bull. Am. Soc. Inform. Sci. Technol. **38**(2), 49–57 (2012)
5. Hoffman, M., Bach, F.R., Blei, D.M.: Online learning for latent dirichlet allocation. In: Advances in Neural Information Processing System (2010)
6. Roy, S.D., Mei, T., Zeng, W., Li, S.: Empowering cross-domain internet media with real-time topic learning from social streams. In: 2012 IEEE International Conference on Multimedia and Expo (ICME), IEEE, pp. 49–54 (2012)
7. Roy, S.D., Mei, T., Zeng, W., Li, S.: SocialTransfer: cross-domain transfer learning from social streams for media applications. In: Proceedings of the 20th ACM International Conference on Multimedia, pp. 649–658 (2012)
8. Roy, S.D., Mei, T., Zeng, W., Li, S.: Towards cross-domain learning for social video popularity prediction. IEEE Trans. Multimedia **15**(6), 1255 (2013)
9. Wall, M.E., Rechtsteiner, A., Rocha, L.M.: Singular value decomposition and principal component analysis. In: Berrar, D.P., Dubitzky, W., Granzow, M. (eds.) A Practical Approach to Microarray Data Analysis, pp. 91–109. Kluwer, Norwell, MA (2003)
10. Bach, F.R., Jordan, M.I.: Learning spectral clustering, with application to speech separation. J. Mach. Learn. Res. **7**, 1963–2001 (2006)
11. Dai, W., Ou, J., Gui-Rong, X., Qiang, Y., Yong, Y.: Eigentransfer: a unified framework for transfer learning. In: Proceedings of the 26th Annual International Conference on Machine Learning, ACM, pp. 193–200 (2009)
12. Lin, F., Cohen, W.W.: Power iteration clustering. In: Proceedings of the 27th International Conference on Machine Learning (ICML-10) (2010)
13. Golub, G.H.: Some modified matrix eigenvalue problems. Siam Rev. **15**(2), 318–334 (1973)
14. Toderici, G., Hrishikesh, A., Marius, P., Luciano, S., Yagnik, J.: Finding meaning on youtube: tag recommendation and category discovery. In: 2010 IEEE Conference on Computer Vision and Pattern Recognition (CVPR), IEEE, pp. 3447–3454 (2010)
15. Mavroeidis, D.: Mind the eigen-gap, or how to accelerate semi-supervised spectral learning algorithms. In: Proceeding of IJCAI, pp. 2692–2697 (2011)

Chapter 2
Learning Social Relations from Videos: Features, Models, and Analytics

Lei Ding and Alper Yilmaz

2.1 Introduction

Despite the progress made during the last decade in video understanding, extracting high-level semantics in the form of relations among the actors in a video is still an under-explored area. This chapter discusses a streamlined methodology to learn interactions between actors, construct social networks, identify communities, and find the leader of each community in a video sequence from a sociological perspective. Specifically, we review one of the first studies reported in [8] toward learning such relations from videos using visual and auditory cues. The main contribution can be stated as the association of low-level video features to social relations by means of machine learning mechanisms, including support vector regression and Gaussian processes. The resulting social network is then analyzed to find communities of actors and identifying the leader of each community, which are two of the most important tasks in social network analytics. Furthermore, as an extension to the basic framework, we discuss the relationship between visual concepts and social relations that has been explored in [9]. In this setting, visual concepts serve as mid-level visual representation in inferring social relations and are compared with features employed in the basic framework.

Recently, researchers have devoted countless efforts on understanding the scene content from video by analyzing the object trajectories and finding common motion patterns [2, 7, 13, 22, 23, 44]. Most of these efforts, however, did not go beyond analyzing or grouping trajectories, or understanding individual actions performed by tracked objects [3, 16, 20, 36, 43]. The computer vision community, generally

L. Ding (✉)
The Ohio State University, Boston, MA 02110, USA
e-mail: leiding326@gmail.com

A. Yilmaz
The Ohio State University, Columbus, OH 43210, USA
e-mail: yilmaz.15@osu.edu

Y. Fu (ed.), *Human-Centered Social Media Analytics*,
DOI: 10.1007/978-3-319-05491-9_2, © Springer International Publishing Switzerland 2014

speaking, did not consider analyzing the video content from a sociological perspective, which would provide systematic understanding of the roles and activities performed by actors based on their relations. In relation to the existing body of work on action or event recognition and analysis, better analyzed social relations, when used with other feature observations, can provide useful contextual information to aid in disambiguating hard-to-recognize actions, events, or objects [29, 38].

In sociology, social structures are believed to be best represented and analyzed using a social network [39]. Social network analytics views social relations in terms of a network consisting of vertices and edges. The vertices represent individual actors within the network, and the edges denote the relations between the actors. The resulting graph structure provides a means to detect and analyze communities in the network. The communities are traditionally detected based on the connectivity between the actors using social network tools, such as the popular modularity algorithm [28]. Social network analytics has recently attracted much interest in the fields of data mining [30, 42] and content analysis of surveillance videos [45].

Due to the availability of visual and auditory information, we chose to perform the proposed techniques on theatrical movies, which contain recordings of social happenings and interactions. In order to address challenges introduced by the generality of relations among movie actors, our approach first aligns the movie script with the frames in the video using closed captions. We note that, the movie script is used only to segment the movie into scenes and provide a basis for generating the scene–actor relation matrix. Alternatively, this information can be obtained using video segmentation and face detection and recognition techniques [4, 46]. A unique characteristic of our proposed framework is its applicability to an adversarial social network, which is a highly recognized but less researched topic in sociology [39], possibly due to the complexity of defining adversarial relations alongside friendship relations. Without loss of generality, an adversarial social network contains two disjoint rival communities composed of actors, where members within a community have friendly relations and across communities have adversarial relations.

In our basic framework, we use visual and auditory information to quantify a grouping cue at the scene level, which serves as soft constraints among the actors. These soft constraints are then integrated to learn interactor affinity. The communities in the resulting social network are discovered by subjecting the interactor affinity matrix to a generalized modularity principle [5]. Each community in a social network typically contains an influential person who strongly connects to all other members of the community. Arguably, we call this influential person as the leader of the community, and detect him/her by adopting eigenvector centrality [33]. An illustration of the communities and their leaders discovered by our approach is given in Fig. 2.1 for the movie titled *G.I. Joe: The Rise of Cobra* (*2009*).

The remainder of the chapter is organized as follows: We start with a brief survey of related work. We then describe our learning-based approach to transform low-level features into grouping cues in Sect. 2.2. Section 2.3 details how we learn social networks from videos. The methodology used to analyze these social networks is described in Sect. 2.4, and is evaluated on a set of videos in Sect. 2.5. Furthermore, the connections between visual concepts and social relations are explored in Sect. 2.6.

Fig. 2.1 Pictorial representation of communities in the movie titled *G.I. Joe: The Rise of Cobra* (*2009*). Our approach automatically detects the two rival communities (*G.I. Joe* and *Cobra*), and identifies their leaders (*Duke* and *McCullen*) visualized as the upscaled characters at the front of each community

Fig. 2.2 Flow diagram of the proposed learning-based framework for constructing social networks

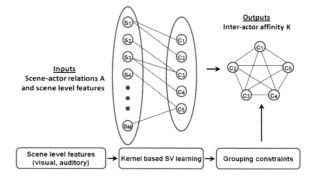

We summarize the chapter in Sect. 2.7. The learning-based framework for constructing social networks is illustrated in Fig. 2.2, for an example video of *M* scenes and 5 interacting actors. The scene–actor relations are visualized in the top row, where a scene is linked to actors that appear in it. A social network representing the video is learned from both scene–actor relations and grouping constraints.

2.1.1 Related Work

Aside from research conducted on social networking in the field of sociology, other fields, which include data mining, computer vision, and multimedia analysis, have been using the ideology behind social networking to solve problems in their respective domains [40, 42, 45]. In this section, we expand our discussion on approaches as they relate to our problem domain.

For detecting communities from surveillance video, a recent study reported in [45] takes advantage of traditional social networking methods. The interactions between the objects are conjectured to occur depending on the proximity heuristic. This heuristic, while far from representing social relations, defines a measurable quantity to define communities in the scene. The authors use traditional modularity [28] to find such communities. Similarly, Ge et al. [18] define the existence of social relations based on the proximity and relative velocity between the objects, which are later used to detect communities in a crowd by means of clustering techniques.

For analysis and segmentation of movies, Weng et al. [40] generate a social network from actor co-occurrence in a scene without attributing them to low-level video features. In their work, the relations are conjectured to be only friendly relations; hence, community can be detected using traditional clustering techniques. While the co-occurrence reasoning resembles our approach, we go one step further and relate them to audiovisual features which are used in a learning framework to define the types of interactions.

Besides, the latest work in computer vision including [15] has strived to identify specific categories of social interaction from video content using Markov random fields (MRFs), but the goal there is not to estimate the social network structure for a group of individuals. Similarly, the study reported in [31] deals with the problem of recognizing social roles played by people in an event via weakly supervised conditional random fields (CRFs). However, the authors did not leverage social network structure for such analysis.

Social network analytics has recently been considered in data mining to find certain interactions within a network generated from log-entries. In two different studies [14, 42] by different groups, the authors have analyzed social networks and their dynamics using Bayesian modeling of social networks and people's interactions. Similar to data mining, in reality mining, researchers have used mobile phone usage to infer social networks using nonvideo data [11, 12]. Due to ubiquitous availability of videos, we believe our framework opens up new directions for studying social phenomena from videos. While it is relatively straightforward to use log-entries or cell phone usage, extracting social content from video presents significant challenges to pattern recognition research.

Broadly speaking, the main difference of our approach from all the papers cited above and many others in the field of sociology is the methodology we have taken to address the social network generation problem. Particularly, existing methods define a heuristic interaction and derive a social network using these heuristics. Before we continue, we would like to draw an analogy between our treatment and the way humans would approach to the same problem. Consider a scene where there are several individuals performing some activities; at first sight, a human observer without knowing the domain and interaction types would immediately consider all individuals are equally related to each other. Based on the duration the observer watches the scene, he will start guessing the type of interactions and derive a social network using his past experience. Similar to human experience, we learn interactions and networks from given training examples and infer social networks in a novel scene. We believe, the proposed approach can benefit other areas in computer vision such as meeting video analysis where modeling individual actions and high-level relations between

Table 2.1 A summary of related work in addition to ours on constructing social networks from data

Data sources	Observed features	Construction techniques	Usability of framework	Examples
From interaction logs	Social interactions	Simple connections	On collected social data, e.g. emails	[14, 42]
From cell phone usage	Call data, etc.	Simple connections	On collections of mobile devices	[11, 12]
From videos (existing)	Tracked people	Proximity heuristics	On surveillance videos	[18, 45]
From videos (this chapter)	Audio-visual cues	Learning approaches	On videos with training labels	[8, 9]

the attendees are important [1, 47]. Finally, some of the aforementioned studies are summarized in Table 2.1, where the novelty and generality of our framework can be readily observed.

2.2 Learning Grouping Cues

Consider a case when the relations among actors have no prior specification and what we observe is only low-level video features. In this setting, communities in the video cannot be explicitly labeled. In our approach, we extract the low-level features from videos, such that kernels on scene-level features provide grouping cues using regression learned from other videos. Unlike many existing approaches, the proposed mapping strategy is data-driven and provides a flexible and extensible approach that can incrementally use new features as they become available.

Before we proceed, let us assume that a video $\mathbb{V}$ is composed of scenes, $s_1, s_2 \ldots s_M$, each of which contains a set of actors and has an associated grouping cue $\beta_i \in [-1, +1]$. The grouping cue serves as a basis to decide whether actors co-occurring in the scene belong to the same community ($\beta_i > 0$) or different communities ($\beta_i < 0$). In our setting, the larger the absolute value of β_i is, the more stringent the corresponding constraints are. In the following discussion, we will detail our approach on estimating such grouping cues from low-level video content.

We conjecture that the interactions among the members of a social community are different from the interactions among the members across different communities. This conjecture imposes a weak grouping assumption due to the fact that we do not need to know the identities of interacting communities. Therefore, labeling from a set of source videos can be generalized to a novel video. In a similar manner, we also conjecture that a video of activities contains low-level features that convey the characteristic types of relationships among the actors performing them. In other words, the relationships between the activities and the actors provide a distinct feature set that can be used to infer if members of a single community or different communities co-occur in the same video segment. For example, boys and girls attending a school interact in distinct ways within and across the groups [19, 27]. Similarly, rivalry across different groups creates adversarial relations when they interact, such as the actions they perform and the words they exchange.

Considering that a scene is the smallest segment in a movie which contains a continued event, low-level features generated from the video and audio of each scene can be used to quantify adversarial and non-adversarial contents. Movie directors often follow certain rules, referred to as the film grammar or cinematic principles in the film literature, to emphasize the adversarial content in scenes. Typically, adversarial scenes contain abrupt changes in visual and auditory contents, whereas these contents change gradually in non-adversarial scenes. Therefore, the visual and auditory features, which quantify adversarial scene content, can be extracted by analyzing the disturbances in the video [32].

In particular for measuring visual disturbance, we follow the cinematic principles and conjecture that for an adversarial scene, the motion field is nearly evenly distributed in all directions (see Fig. 2.3 for illustration). For generating the optical flow distributions, we use the Kanade–Lucas–Tomasi tracker [34] within the scene bounds and use good features to track. Alternatively, one can use dense flow field generated by estimating optical flow at each pixel [26]. The visual disturbance in the observed flow field can be measured by entropy of the orientation distribution as shown in Fig. 2.4. Specifically, we apply a moving window of 10 frames with 5 frames overlapping in the video for constructing the orientation histograms of optical flows. We use histograms of optical flow vectors weighted by the magnitude of motion. The number of orientation bins is set to 10 and the number of entropy bins in the final feature vector is set to 5. As can be observed in Fig. 2.5, flow distributions generated from adversarial scenes tend to be uniformly distributed and thus, they consistently have more high-entropy peaks compared to non-adversarial scenes. This observation serves as the basis for distinguishing the two types of scenes.

Auditory features extracted from the accompanying movie audio are used together with the visual features to improve the performance. We adopt a combination of temporal and spectral auditory features discussed in [24, 32], which are energy peak ratio, energy entropy, short-time energy, spectral flux, and zero crossing rate:

- Energy peak ratio $EPR = \frac{p}{S}$, where p is the number of energy peaks and S is length of an audio frame;
- Energy entropy $EE = -\sum_{i=1}^{K} e_i \log e_i$, where a audio frame is divided into K sub-windows. For sub-window i, energy e_i is computed;
- Short-time energy $SE = \sum_{i=1}^{S} x_i^2$, where S is the length of an audio frame;
- Spectral flux $SF = \frac{1}{KF} \sum_{i=2}^{K} \sum_{j=1}^{F} (\varepsilon_{i,j} - \varepsilon_{i-1,j})^2$, where $\varepsilon_{i,j}$ is the spectral energy at sub-window i and frequency channel j;
- Zero crossing rate $ZCR = \frac{1}{2S} \sum_{i=1}^{S} |sgn(x_i) - sgn(x_{i-1})|$, where sgn stands for a sign function.

Specifically, these features are computed for sliding audio frames that are 400 ms in length. The means of these features over the duration of the scene constitute

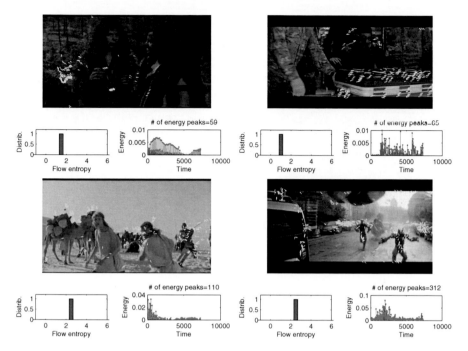

Fig. 2.3 Visual and auditory characteristics of adversarial scenes. *Top row* non-adversarial scenes from *Year One (2009)* and *G.I. Joe: The Rise of Cobra (2009)*; *Bottom row* adversarial scenes from these two movies. Optical flow vectors are superimposed on the frames and computed features are shown as plots for a temporal window of 10 video frames, including entropy distribution of optical flow vectors and detected energy peaks (*red* dots in energy signals). *Note* For color interpretation see online version

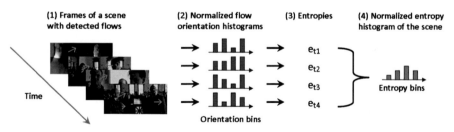

Fig. 2.4 Generation of the normalized entropy histogram from orientation distributions of optical flows detected from a scene

a feature vector. A sample auditory feature (energy peaks) is shown in Fig. 2.3 for both adversarial and non-adversarial scenes. It can be observed that adversarial scenes have more peaks in energy signals, which are moving averages of squared audio signals.

The visual and auditory features provide two vectors per scene (5 dimensional visual and 5 dimensional auditory), which are used to estimate a real-valued group-

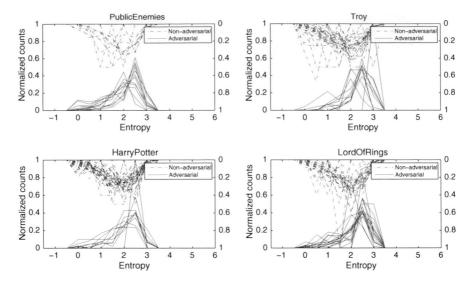

Fig. 2.5 Visualization of entropy histogram feature vectors extracted from four example movies. The two classes (adversarial and non-adversarial) have distinct patterns, in that adversarial scenes tend to consistently produce strong peaks in high entropies. Best viewed in color

ing cue $\beta_i \in [-1, +1]$ of the scene s_i. To achieve this goal, we use support vector regression (SVR) [35], which has been successfully used to solve various problems in computer vision literature [10, 21]. We apply a radial basis function to both the visual and auditory feature vectors, which leads to two kernel matrices $\mathcal{K}_v$ and $\mathcal{K}_a$, respectively. The two kernel bandwidths can be chosen by using cross-validation. The joint kernel is then computed as the multiplication kernel: $\hat{\mathcal{K}}(u, v) = \mathcal{K}_v(u, v)\mathcal{K}_a(u, v)$. In support vector regression, the goal is to find a function $g(\cdot)$ that has at most ε deviation from the labeled targets for all the training data, and at the same time is as flat as possible. It is shown that the final decision function can be written as:

$$\beta_i = g(s_i) = \sum_{j=1}^{L}(\alpha_j - \alpha_j^*)\hat{\mathcal{K}}_{l_j,i} + b, \tag{2.1}$$

where the coefficient b is offset, α_i and α_i^* are the Lagrange multipliers for labeling constraints, L is the number of labeled examples, and l_j is the index for the jth labeled example.

In our problem domain, the joint kernel together with training video scenes and their grouping cues $\beta_i = +1$ (scene with members of only one community) and $\beta_i = -1$ (scene with members from different communities) leads to grouping constraints for a novel video. This is achieved by estimating the corresponding grouping cues β_i using the regression learned from labeled video scene examples from other videos in the training set.

2.3 Learning Social Networks

Consider actors co-occurring in a video. We conjecture that these actors co-occur more often if they are members of the same community. The higher the number of co-occurrences for the same-community members is, the more positive grouping cues are present compared to the negative ones. The combination of these two factors plays a significant role in our social network learning methodology. In the following discussion, we will first describe the representations we use which is followed by how social relations are learned.

2.3.1 Basic Representations

The temporal occurrence of an actor c_i in a video is represented by a boolean appearance function $\psi_i : T \rightarrow \{0, 1\}$, where the duration of a video $T \subset \mathbb{R}^+$. In practice, we only have access to its sampled version. Suppose that the sampling period is of length t seconds. According to the Nyquist's sampling theorem, as long as $t \leq \min_i\{1/2B_i\}$, where B_i is the highest frequency (in hertz) of actor i's appearance function, the continuous appearance information and their co-occurrence can be determined from those discrete samples.

A video $\mathbb{V}$ is composed of nonoverlapping M scenes, where each scene s_i contains interactions among a set of actors. In order for computational convenience, the appearance functions of actors are approximated as a scene–actor relation matrix denoted by $A = \{A_{i,j}\}$, where $A_{i,j} = 1$ if there exists $t \in L_i$, where L_i is the temporal interval of s_i, such that $\psi_j(t) = 1$. This, for a movie, can be obtained by searching for speaker names in the script. This representation is reminiscent of the actor–event graph in social network analysis [39]. While the actor relations in A can be directly used for construction of the social network, we will demonstrate later that the use of visual and auditory scene features can lead to a better social network representation.

Finally, the social network is represented as an undirected graph $G(V, E)$ with cardinality $|V|$. In this graph, the vertices represent the actors

$$V = \{v_i : \text{vertex } v_i \sim \text{actor } c_i\}, \tag{2.2}$$

and the edges define the interactions between the actors

$$E = \{(v_i, v_j)|v_i, v_j \in V\}. \tag{2.3}$$

The resulting graph G is a fully connected graph with an affinity matrix K of size $|V| \times |V|$. The entry in the affinity matrix $K(c_i, c_j)$ for two actors c_i and c_j is a real-valued weight, which is decided by an affinity learning method that will be introduced next in this section. The values in the affinity matrix serve as the basis for social network analytics, including determining communities and leaders in the social network.

2.3.2 Actor Interaction Model

Let c_i be actor i, and $\mathbf{f} = (f_1, \ldots, f_N)^T$ be the vector of community memberships containing ± 1 values, where f_i refers to the membership of c_i. Let $\mathbf{f}$ distribute according to a zero-mean identity-covariance Gaussian process

$$P(\mathbf{f}) = (2\pi)^{-N/2} \exp^{-\frac{1}{2}\mathbf{f}^T\mathbf{f}} . \tag{2.4}$$

In order to model the information contained in the scene–actor relation matrix A and the aforementioned grouping cue of each scene β_i, we assume the following distributions:

- if c_i and c_j occur in a non-adversarial scene k ($\beta_k \geq 0$), we assume $f_i - f_j \sim \mathcal{N}(0, \frac{1}{\beta_k^2})$;
- if c_i and c_j occur in an adversarial scene k ($\beta_k < 0$), we assume $f_i + f_j \sim \mathcal{N}(0, \frac{1}{\beta_k^2})$.

Therefore, if $\beta_i = 0$, then the constraint imposed by a scene becomes inconsequential, which corresponds to the least confidence in the constraint. On the other hand, if $\beta_i = \pm 1$, the corresponding constraint becomes the strongest. Because of the distributions we use, none of the constraints is hard, making our method relatively flexible and insensitive to prediction errors. Applying the Bayes' rule, the posterior probability of $\mathbf{f}$ given the constraints is defined in a continuous formulation as the following:

$$
\begin{aligned}
P(\mathbf{f}|\{\psi_k\}, \beta) = P(\mathbf{f}) \exp &\left\{ -\sum_{i,j} \int_{t \in \{t:\beta(t)\geq 0\}} \psi_i(t)\psi_j(t) \frac{(f_i - f_j)^2 \beta(t)^2}{2} dt \right. \\
&\left. -\sum_{i,j} \int_{t \in \{t:\beta(t)<0\}} \psi_i(t)\psi_j(t) \frac{(f_i + f_j)^2 \beta(t)^2}{2} dt \right\} \\
\propto \exp &\left\{ -\frac{1}{2}\mathbf{f}^T\mathbf{f} - \sum_{i,j} \int_{t \in \{t:\beta(t)\geq 0\}} \psi_i(t)\psi_j(t) \frac{(f_i - f_j)^2 \beta(t)^2}{2} dt \right. \\
&\left. -\sum_{i,j} \int_{t \in \{t:\beta(t)<0\}} \psi_i(t)\psi_j(t) \frac{(f_i + f_j)^2 \beta(t)^2}{2} dt \right\} .
\end{aligned}
\tag{2.5}
$$

Translating this equation into its discrete version, we have:

$$P(\mathbf{f}|A, \beta) = P(\mathbf{f}) \, \Pi_{k:\beta_k \geq 0} \Pi_{c_i,c_j \in s_k} \exp \frac{-(f_i - f_j)^2 \beta_k^2}{2}$$

$$\Pi_{k:\beta_k < 0} \Pi_{c_i,c_j \in s_k} \exp \frac{-(f_i + f_j)^2 \beta_k^2}{2}$$

$$\propto \exp \left\{ -\frac{1}{2} \mathbf{f}^T \mathbf{f} - \sum_{k:\beta_k \geq 0} \sum_{c_i,c_j \in s_k} \frac{(f_i - f_j)^2 \beta_k^2}{2} \right.$$

$$\left. - \sum_{k:\beta_k < 0} \sum_{c_i,c_j \in s_k} \frac{(f_i + f_j)^2 \beta_k^2}{2} \right\}. \tag{2.6}$$

It can be verified that $P(\mathbf{f}|A, \beta) \propto \exp(-\frac{1}{2}\mathbf{f}^T K^{-1}\mathbf{f})$ is a Gaussian process with zero mean. Using $K_{i,j} = E\{f_i f_j | A, \beta\}$ as the learned affinity between c_i and c_j, it follows that $K = M^{-1}$, where

$$M_{i,j} = \begin{cases} \sum_{k:c_i,c_j \in s_k, \beta_k < 0} \beta_k^2 - \sum_{k:c_i,c_j \in s_k, \beta_k \geq 0} \beta_k^2 & \text{if } i \neq j \\ 1 + \sum_{l \neq i} \sum_{k:c_i, c_l \in s_k} \beta_k^2 & \text{if } i = j \end{cases}. \tag{2.7}$$

The resulting K is symmetric and positive definite. However, unlike an affinity matrix from a Gaussian kernel, it may contain negative values. The proposed approach has two special cases:

- In the case when $\beta_i = 1$, then the aforementioned learning mechanism reduces to a co-occurrence-based approach which is a traditional tool in social network analysis [5, 28]. Specifically, $M_{i,j}$, for $i \neq j$, represents the minus value of the number of scenes where c_i and c_j occur together. This reduced scheme does not utilize the video/audio feature-based prediction of grouping cues, and serves as a natural baseline in this chapter.
- If we use fixed variance parameters in the assumed distributions instead of the learned ones, our affinity learning method reduces to the affinity propagation approach proposed in [25].

2.4 Social Network Analytics

A primary goal of social network analytics is finding groups of actors to form social communities and detecting the most influential actor within a community, which we arguably refer to as the leader of a community. Traditionally, communities are detected using spectral clustering techniques tailor-made for social settings, such as the popular modularity-cut algorithm [28]. A recent study reported in [5] has shown that the performance of modularity cut can be increased by introducing a generalized objective referred to as the max–min modularity. The max–min modularity clustering, however, assumes unweighted edges and is not directly suitable for our social networks which contain learned weighted edges.

In our design, we first generate a principal affinity matrix K' by the following rules: $K'_{i,j} = K_{i,j}$ for $K_{i,j} > 0$, and $K'_{i,j} = 0$ for other entries. We then generate a complementary affinity matrix K'' by the following rules: $K''_{i,j} = -K_{i,j}$ for $K_{i,j} < 0$, and $K''_{i,j} = 0$ for other entries. The matrix K'' represents the unrelatedness between vertices in the network in terms of community memberships. Adopting the strategy in [5] and using K' and K'', we formulate the max–min modularity criterion as $Q_{MM} = Q_{\max} - Q_{\min}$ for:

$$Q_{\max} = \frac{1}{2m'} \sum_{i,j} \left(K'_{ij} - \frac{k'_i k'_j}{2m'} \right) (f_i f_j + 1) \triangleq \frac{1}{2m'} \sum_{i,j} B'_{i,j}(f_i f_j + 1), \quad (2.8)$$

$$Q_{\min} = \frac{1}{2m''} \sum_{i,j} \left(K''_{ij} - \frac{k''_i k''_j}{2m''} \right) (f_i f_j + 1) \triangleq \frac{1}{2m''} \sum_{i,j} B''_{i,j}(f_i f_j + 1), \quad (2.9)$$

where $m' = \frac{1}{2} \sum_{ij} K'_{ij}$, $k'_i = \sum_j K'_{ij}$, $m'' = \frac{1}{2} \sum_{ij} K''_{ij}$, $k''_i = \sum_j K''_{ij}$ and the term $\frac{k'_i k'_j}{2m'}$ represents the expected edge strength between the actors c_i and c_j [28]. Based on this observation, we note that $K'_{i,j} - \frac{k'_i k'_j}{2m}$ measures how much the connection between two actors is stronger than what would be expected between them, and serves as the basis for keeping the two actors in the same community. In this formulation, the max–min modularity Q_{MM} roots from the conditions for a good network division that (1) edge strength across communities should be smaller than expected, and (2) unrelated actors within a community should be minimal. These conditions can be realized by maximizing Q_{MM}. Using standard eigenanalysis, it follows that the eigenvector $\mathbf{u}$ of $\frac{1}{2m'} B' - \frac{1}{2m''} B''$ with the largest eigenvalue maximizes a relaxed version of Q_{MM}. The resulting eigenvector solution contains real values, and we threshold them at the 0 level to obtain the desired community memberships. That is, we let $f_i = +1$ if $u_i \geq 0$, and $f_i = -1$ if otherwise.

Once the communities in the video are extracted, their leaders are detected by analyzing the centrality of each actor in the community. In the sociology literature, the centrality of an actor is traditionally defined by its degree or betweenness [17]. In this chapter, we, rather, adopt a new technique which is referred to as the eigen-centrality [33] due to its relation to proposed community detection approach. Let the centrality score, x_i for the ith actor be proportional to the sum of the scores of all vertices which are connected to it: $x_i = \frac{1}{\lambda} \sum_{j=1}^{N} K'_{i,j} x_j$, where N is the total number of actors in the video and λ is a constant. It follows from this notation that the centralities of actors satisfy $K'\mathbf{x} = \lambda \mathbf{x}$ in the vector form. It can be shown that the eigenvector with largest eigenvalue provides the desired centrality measure [33]. Therefore, if we let the eigenvector of K' with the largest eigenvalue be $\mathbf{v}$, the leaders of the two communities are given by $\arg\max_{i:u_i \geq 0} v_i$ and $\arg\max_{i:u_i < 0} v_i$, respectively. In our problem domain, when the communities correspond to two adversarial social groups, their expected leaders relate to the hero or the villain in the video.

2.5 Experiments

For qualitative and quantitative evaluation of the proposed approach, we generate a dataset of 10 movies which contains recent and classical theatrical movies that cover a range of genres including action, adventure, fantasy, and drama.[1] The movies in our dataset broadly contain two rival communities with a designated leader for each community. For each movie with statistics tabulated in Table 2.2, the dataset contains visual and auditory features, movie script, and closed caption data, all of which are temporally aligned.

For movie domain, in order to align visual and auditory features with the script, we require temporal segmentation of the movie into scenes, which provides start and stop timings for each scene. This segmentation process is guided by the accompanying movie script and closed captions. The script is usually a draft version with no time tagging and lacks professional editing, while the closed captions are composed of lines d_i, which contain timed sentences uttered by actors. The approach we use to perform this task can be considered as a variant of the alignment technique in [6]:

1. Divide the script into scenes, each of which is denoted as s_i. Similarly, closed captions are divided into lines d_i.
2. Define $\mathscr{C}$ to be a cost matrix. Compute the percentage p of the words in closed caption d_j matched with scene s_i while respecting the order of words. Set the cost as $\mathscr{C}_{i,j} = 1 - p$.
3. Apply dynamic time warping to $\mathscr{C}$ for estimating start t_1^i and stop times t_2^i of s_i, which respectively correspond to the smallest and largest time stamps for closed captions matched with s_i.

At the end of this process, we generate the start and stop timings of all scenes. Due to the fact that publicly available scripts for movies are not perfectly edited, the temporal segmentation may not be precise. Nevertheless, our approach is robust to such inaccuracies in segment boundaries.

In the following discussion, we analyze social networks with accompanying affinity matrices generated from

- the actor co-occurrence information reflected in matrix A (co-occurrence), which is more traditional in sociology;

---

[1] The movies in or dataset are (1) *G.I. Joe: The Rise of Cobra (2009)*; (2) *Harry Potter and the Half-Blood Prince (2009)*; (3) *Public Enemies (2009)*; (4) *Troy (2004)*; (5) *Braveheart (1995)*; (6) *Year One (2009)*; (7) *Coraline (2009)*; (8) *True Lies (1994)*; (9) *The Chronicles of Narnia: The Lion, the Witch and the Wardrobe (2005)*; and (10) *The Lord of the Rings: The Return of the King (2003)* .

Table 2.2 Statistics of movies in our dataset which includes the number of scenes in the movie, the number of lines in closed caption data, the total number of actors in the movie, and the number of actors in one of the two communities

Movies enumerated in footnote 1	(1)	(2)	(3)	(4)	(5)	(6)	(7)	(8)	(9)	(10)
# of scenes	198	151	238	226	116	51	105	297	188	199
# of captions in lines	1,143	1,585	1,063	1,155	1,337	1,515	1,293	1,262	1,099	1,402
# of actors in total	11	7	10	10	7	7	6	8	10	9
# of actors in community 1	6	4	6	5	3	3	4	6	7	7

Table 2.3 Community leaders discovered using the proposed framework

Movies	(1)	(2)	(3)	(4)	(5)
Community 1	Hawk	Harry	Dillinger	Achilles	MacClan.
Community 2	McCullen	Snape	Purvis	Androm.	Longsha.
Movies	(6)	(7)	(8)	(9)	(10)
Community 1	Zed	Coraline	Trilby	Susan	Frodo
Community 2	Abraham	OtherMo.	Salim	Witch	WitchKing

The names in bold face refer to correct ones, whereas those in italics are not

- in addition to co-occurrence, scene-level grouping cues β_i which are learned from video and audio contents using the proposed approach.

In order to evaluate the contribution of these features, we provide comparisons of collective use of visual and auditory features with their individual use in extraction of communities and their leaders. In Fig. 2.6, we show graphical representations of the social networks for 10 movies learned from both visual and auditory features using the proposed approach. The color codes in the figure reflect the strength of affinity between actors. We observe that intercommunity connections tend to be weaker than certain intracommunity ones.

The affinity between the actors is strongly related to the grouping cues of the scenes in which they appear. This relation suggests validation of how effective the support vector regression (SVR) is for estimating the grouping cues. In order to facilitate this, we compute the mean square error (MSE) as our error measure over all the scenes in each movie, and average the resulting MSEs over all 10 movies. When both the visual and auditory features are used, the MSE is estimated as 0.61. In contrast, when only one of the features is used MSE increases to 0.80 for visual only and 0.77 for auditory only. These numbers translate into accuracy rates for

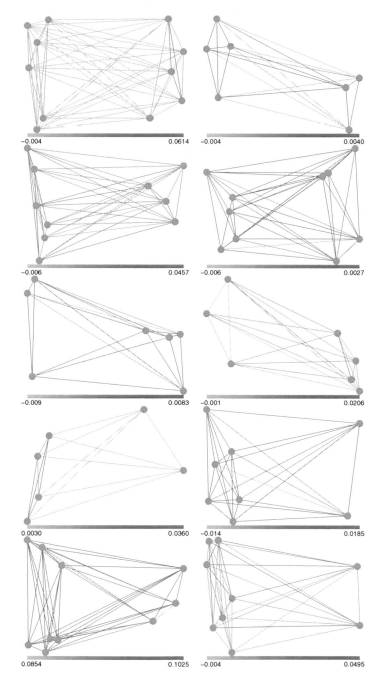

Fig. 2.6 Social networks generated using the proposed approach for the 10 movies in our dataset. Actors (vertices) are placed on the *left* and *right* with respect to communities they belong to. The strength of affinity is indicated by *pinkness* of the edges: the stronger the edge is the *pinker* it is. Best viewed in color. *Note* For color interpretation see online version

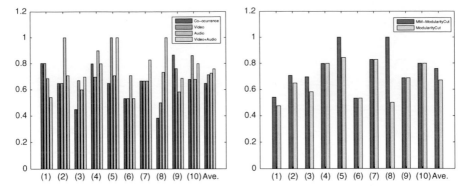

Fig. 2.7 Accuracy of social network analytics in F_1 measures. *Left*: comparison of four approaches, where the proposed one is video+audio; *Right*: comparison of two modularity algorithms (max–min vs. original), with the proposed video+audio approach

predicting if a scene is adversarial or non-adversarial. Respectively, the accuracy rates are computed as 81.6, 78.2, and 78.7 % for collective feature use, visual only and auditory only. These numbers reflect that the grouping cue estimates of scenes can be further utilized to infer the relations among the movie actors.

The accuracy of community detection relates to how precise the assignment of the actors is into each one of the community. Considering that a community is a set of actors, the accuracy can be measured using the precision and recall values of predicted assignments given the ground truth. For each community these two values can be combined into an F_1 measure, which is the harmonic mean of precision and recall. This measure takes into account the possible imbalance in the size of communities and has been widely adopted. Considering that the movies in our dataset contains more than one community, we report the average F_1 measure over detected communities as the final detection accuracy for each movie.

From the quantitative evaluations shown in Fig. 2.7, for four movies visual features help enhance performance appreciably. Overall, auditory features improve the performance slightly more than the visual features when they are used independently. Their combination, however, provides the best performance, which on the average leads to an F_1 measure of 76.0 %. This score, when compared to using only the actor co-occurrence to generate the social network, improves the grouping performance by 11.1 %. In the same figure, we also show that the modified max–min modularity, when compared to the traditional modularity computed from K, improves the F_1 measure by 8.9 %.

As discussed in Sect. 2.4, the community assignment of actors is realized by analyzing the eigenspace of $\frac{1}{2m'} B' - \frac{1}{2m''} B''$. In order to visualize this assignment process, we map the actors in the movie into coordinates defined by the two eigenvectors with highest eigenvalues. This mapping provides an optimal way to visualize the interactor relations in two dimensions. In Fig. 2.8, we illustrate the ground truth

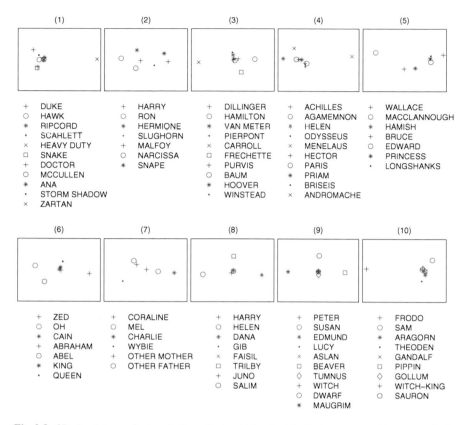

(1)	(2)	(3)	(4)	(5)
+ DUKE	+ HARRY	+ DILLINGER	+ ACHILLES	+ WALLACE
○ HAWK	○ RON	○ HAMILTON	○ AGAMEMNON	○ MACCLANNOUGH
* RIPCORD	* HERMIONE	* VAN METER	* HELEN	* HAMISH
· SCARLETT	· SLUGHORN	· PIERPONT	· ODYSSEUS	+ BRUCE
× HEAVY DUTY	+ MALFOY	× CARROLL	× MENELAUS	○ EDWARD
□ SNAKE	○ NARCISSA	□ FRECHETTE	+ HECTOR	* PRINCESS
+ DOCTOR	* SNAPE	+ PURVIS	○ PARIS	· LONGSHANKS
○ MCCULLEN		○ BAUM	* PRIAM	
* ANA		* HOOVER	· BRISEIS	
· STORM SHADOW		· WINSTEAD	× ANDROMACHE	
× ZARTAN				

(6)	(7)	(8)	(9)	(10)
+ ZED	+ CORALINE	+ HARRY	+ PETER	+ FRODO
○ OH	○ MEL	○ HELEN	○ SUSAN	○ SAM
* CAIN	* CHARLIE	* DANA	* EDMUND	* ARAGORN
+ ABRAHAM	+ WYBIE	· GIB	· LUCY	· THEODEN
○ ABEL	+ OTHER MOTHER	× FAISIL	× ASLAN	× GANDALF
* KING	○ OTHER FATHER	□ TRILBY	□ BEAVER	□ PIPPIN
· QUEEN		+ JUNO	◇ TUMNUS	◇ GOLLUM
		○ SALIM	+ WITCH	+ WITCH-KING
			○ DWARF	○ SAURON
			* MAUGRIM	

Fig. 2.8 2D visual maps of actor relations. *Red* and *blue* stand for the two communities, respectively, according to our ground truth labeling. Best viewed in color. *Note* For color interpretation see online version

in red and blue colors, respectively, for the two communities.[2] As can be observed, the actors who belong to separate communities tend to lie apart.

As we mentioned earlier, the eigenvector of K' with the highest eigenvalue provides the leaders of communities. In Table 2.3, we tabulate these leaders with their pictures for the two rival communities for each movie. The predicted leaders who correspond to the true leaders in the movie are shown in bold face, while incorrect leaders are shown in italics. Overall, it can be observed that many of the leaders are successfully discovered by our framework.[3]

---

[2] In movie (10), *Gollum* has a good personality except for when he is close to the ring. The ring changes the good behavior of the actors to bad except for *Frodo*.

[3] Ground truth leaders are: (1) Duke and McCullen; (2) Harry and Snape; (3) Dillinger and Purvis; (4) Achilles and Hector; (5) Wallace and Longshanks; (6) Zed and King; (7) Coraline and Other Mother; (8) Harry and Salim; (9) Aslan and Witch; and (10) Frodo and Witch-king.

2.6 Using Visual Concepts

In this section, we discuss an extension to the basic framework of learning social relations from videos. Visual concepts, such as "beach", "cheering", and "shooting", are detected from the video content are employed as mid-level representation, which is used as the basis for inferring social relations. The main intuition behind this approach is that visual concept detection, compared to low-level video information, provides useful semantic features for inferring social relations. For example, individuals involved in a fighting scene tend to be enemies, while individuals jogging leisurely together tend to be friends. To this end, we leverage support vector machine (SVM) detectors for 374 trained visual concepts provided in [41], due to its public availability and broadness.

For detecting base concepts, we use the three low-level features exploited in [41], which include the color, texture, and edges computed from keyframes in a video. In specific, for the grid color moment (GCM) feature, we extract the first 3 moments of the 3 channels in the CIE LUV color space over 5×5 fixed grid partitions, and aggregate the features into a single 225-dimensional feature vector. The texture is modeled by the Gabor texture (GT) feature, which we extract by taking 4 scales and 6 orientations of Gabor transformations and use their means and standard deviations. This process provides a texture feature in the form of a vector with 48 dimensions. The edge content in the image is modeled using the edge direction histogram (EDH), which is composed of 73 dimensions corresponding to 72 bins of edge direction quantized at 5 degree intervals and 1 bin for non-edge points.

We use the temporal extent of a scene to extract observations relating to the social content and consider that the visual content in a scene is represented by its keyframes. Following the extraction of keyframes, we compute low-level features for each keyframe i. These features are then used to estimate a normalized score, which is the raw decision value from an SVM transformed by a logistic function, from the SVMs which are independently trained on each low-level feature: $f_{i,j}$ for three feature types $j = 1, 2, 3$. The average of the three scores $f_i = \frac{1}{3}(f_{i,1} + f_{i,2} + f_{i,3})$ is used as the overall score for the ith keyframe. Finally, we compute the visual concept score by max-pooling over all keyframes within the scene bounds, $\max_i \overline{f}_i$. This process, when performed for all 374 visual concepts, results in a 374-dimensional semantic vector representing the scene. Each element of the semantic vector provides the confidence score corresponding to a semantic concept. Since previous research has suggested that grid- and global-based feature representations extracted from keyframes and SVM classification lead to strong concept detection systems [41], this mid-level representation is used as the basis for inferring social relations.

It is clear that not all the dimensions of the semantic vectors are equally informative toward social relations. Besides, detecting some of the visual concepts may be unsatisfactory due to their large variability or relatively small spatiotemporal extents. To address this issue, we employ a supervised dimension reduction method known as kernel local Fisher discriminant analysis (KLFDA) [37], which has an analytic form of the embedding transformation. By applying this data-dependent transform,

Table 2.4 Comparative analysis of features for social relational learning

Methods	Prec(+)	Prec(-)	Prec(ave) (%)	F_1 (%)
Baseline features	–	–	78.2	76.0
Visual concepts (d=50)	83.1 %	85.0 %	84.1	81.9

The measures used are as follows: Prec(+): precision of β estimates for the positive class; Prec(-): precision of β estimates for the negative class; Prec(ave): average precision; and F_1 measure for community detection averaged over all videos

we derive a more informative and compact representation, which is a d-dimensional vector for each video scene, for learning the social relations.

Following Sect. 2.2, we assume that a video is composed of M scenes, $s_1, s_2 \ldots s_M$, each of which contains a set of actors and has an associated grouping cue β_i. The grouping cue serves as a basis to decide whether the actors co-occurring in a scene belong to the same or different communities. To estimate the grouping cues β_i from the d-dimensional transformed semantic vectors, we use support vector regression, with a radial basis function kernel over the d-dimensional transformed concept score vectors. On the same 10-movie dataset, we are able to compare the performance of using visual concepts versus the other approach in estimating grouping cues. Quantitative analysis tabulated in Table 2.4 is performed on the baseline features detailed in Sect. 2.2 versus the visual concepts described in this section. It should be noted that the F_1 score associated with visual concepts is partially accounted for by the commnunity detection method in [9]. However, it can be seen that visual concept-based features work better than the baseline features by a large margin in the crucial task of grouping cue estimation, with no use of tailored visual features toward the "adverseness" of a scene. Besides, the generality of visual concepts for social relational learning is confirmed by a larger scale study, such as that in [9].

2.7 Summary

In this chapter, we have presented a framework for learning the relations among actors from videos using a social network approach. We have used visual and auditory features to characterize the grouping cues. By using an affinity learning procedure, we incorporate these grouping cues, and make informed decisions in constructing and analyzing the corresponding social network. Extensive analysis on a set of videos has validated the effectiveness of our framework in high-level understanding of social interactions. Besides, leveraging visual concepts has also been considered as mid-level representation for inferring social relations.

Further, it is natural to apply our framework to other problem domains, such as surveillance videos, where behaviors of interest can be related to the interactions between objects in a scene, and meeting videos, where people's modes of interaction may be related to their social groups or organizations. In such settings, it may be possible to include recognizable human actions in generating grouping cues. The proposed framework also contributes to sociology, in that our framework can aid sociological discovery by automatically extracting communities from videos, given available group labeling of the desired type. Uncovering and quantifying such patterns of generalizability may be of interest to sociologists.

References

1. Al-Hames, M., Lenz, C., Reiter, S., Schenk, J., Wallhoff, F., Rigoll, G.: Robust multi-modal group action recognition in meetings from disturbed videos with the asynchronous hidden markov model. In: International Conference on Image Processing (2007)
2. Ali, S., Basharat, A., Shah. M.: Chaotic invariants for human action recognition. In: IEEE International Conference on Computer Vision (2007)
3. Alon, J., Athitsos, V., Yuan, Q., Sclaroff, S.: A unified framework for gesture recognition and spatiotemporal gesture segmentation. IEEE Trans. Pattern Anal. Mach. Intell. **31**(9), 1685–1699 (2009)
4. Arandjelović, O., Zisserman, A.: Automatic face recognition for film character retrieval in feature-length films. In: ACM International Conference on Image and Video Retrieval (2005)
5. Chen, J., Zaiane, O., Goebel, R.: Detecting communities in social networks using max-min modularity. In: SIAM Conference on Data Mining (2009)
6. Cour, T., Jordan, C., Miltsakaki, E., Taskar, B.: Movie/script: alignment and parsing of video and text transcription. In: European Conference on Computer Vision (2008)
7. Ding, L., Fan, Q., Hsiao, J., Pankanti, S.: Graph based event detection from realistic videos using weak feature correspondence. In: International Conference on Acoustics, Speech, and Signal Processing (2010)
8. Ding, L., Yilmaz, A.: Learning relations among movie characters: a social network perspective. In: European Conference on Computer Vision (2010)
9. Ding, L., Yilmaz, A.: Inferring social relations from visual concepts. In: International Conference on Computer Vision (2011)
10. Dufrenois, F., Colliez, J., Hamad, D.: Crisp weighted support vector regression for robust single model estimation: application to object tracking in image sequences. In: IEEE Conference on Computer Vision and Pattern Recognition (2007)
11. Eagle, N., Pentland, A.: Eigenbehaviors: identifying structure in routine. Behav. Ecol. Sociobiol. **63**(7), 1057–1066 (2009)
12. Eagle, N., Pentland, A., Lazer, D.: Inferring social network structure using mobile phone data. Proc. Nat. Acad. Sci. **106**(36), 15274–15278 (2009)
13. Efros, A.A., Berg, A.C., Mori, G., Malik, J.: Recognizing action at a distance. In: IEEE International Conference on Computer Vision (2003)
14. Fan, Y., Shelton, C.R.: Learning continuous-time social network dynamics. In: Conference on Uncertainty in Artificial Intelligence (2009)
15. Fathi, A., Hodgins, J.K., Rehg, J.M.: Social interactions: a first-person perspective. In: IEEE Conference on Computer Vision and Pattern Recognition (2012)
16. Fathi, A., Mori, G.: Action recognition by learning mid-level motion features. In: IEEE Conference on Computer Vision and Pattern Recognition (2008)
17. Freeman, L.: Centrality in social networks: conceptual clarification. Soc. Netw. **1**(3), 215–239 (1979)
18. Ge, W., Collins, R., Ruback, B.: Automatically detecting the small group structure of a crowd. In: IEEE Workshop on Applications of Computer Vision (2009)
19. Holden, C.: Giving girls a chance: patterns of talk in co-operative group work. Gend. Educ. **5**(2), 179–189 (1993)

20. Jiang, H., Fels, S., Little, H.: A linear programming approach for multiple object tracking. In: IEEE Conference on Computer Vision and Pattern Recognition (2007)
21. Kusakunniran, W., Wu, Q., Zhang, J., Li, H.: Support vector regression for multi-view gait recognition based on local motion feature selection. In: IEEE Conference on Computer Vision and Pattern Recognition (2010)
22. Kyriazis, N., Argyros., A.: Physically plausible 3d scene tracking: the single actor hypothesis. In: IEEE Conference on Computer Vision and Pattern Recognition (2013)
23. Laptev, I., Lindeberg, T.: Space-time interest points. In: IEEE International Conference on Computer Vision (2003)
24. Lin, J., Wang, W.: Weakly-supervised violence detection in movies with audio and video based co-training. In: Pacific-Rim Conference on Multimedia (2009)
25. Lu, Z., Carreira-Perpinan, M.A.: Constrained spectral clustering through affinity propagation. In: IEEE Conference on Computer Vision and Pattern Recognition (2008)
26. Lucas, B.D., Kanade, T.: An iterative image registration technique with an application to stereo vision. In: International Joint Conferences on Artificial Intelligence (1981)
27. Myhill, D.: Bad boys and good girls? patterns of interaction and response in whole class teaching. Br. Educ. Res. J. **28**(3), 339–352 (2002)
28. Newman, M.E.J.: Modularity and community structure in networks. Proc. Nat. Acad. Sci. **103**(23), 8577–8582 (2006)
29. Pei, M., Dong, Z., Zhao, M.: Event recognition based on social roles in continuous video. In: IEEE International Conference on Multimedia and Expo (2013)
30. Qiu, J., Lin, Z., Tang, C., Qiao, S.: Discovering organizational structure in dynamic social network. In: IEEE International Conference on Data Mining (2009)
31. Ramanathan, V., Yao, B., Fei-Fei, L.: Social role discovery in human events. In: IEEE Conference on Computer Vision and Pattern Recognition (2013)
32. Rasheed, Z., Shah, M.: Movie genre classification by exploiting audio-visual features of previews. In: International Conference on Pattern Recognition (2002)
33. Ruhnau, B.: Eigenvector-centrality? a node-centrality. Soc. Netw. **22**(4), 357–365 (2000)
34. Shi, J., Tomasi, C.: Good features to track. In: IEEE Conference on Computer Vision and Pattern Recognition (1994)
35. Smola, A.J., Schölkopf, B.: A tutorial on support vector regression. Stat. Comput. **14**(3), 199–222 (2004)
36. Song, Y., Morency, L.-P., Davis, R.: Action recognition by hierarchical sequence summarization. In: IEEE Conference on Computer Vision and Pattern Recognition (2013)
37. Sugiyama, M.: Dimensionality reduction of multimodal labeled data by local Fisher discriminant analysis. J. Mach. Learn. Res. **8**, 1027–1061 (2007)
38. Wang, G., Gallagher, A., Luo, J., Forsyth, D.: Seeing people in social context: recognizing people and social relationships. In: European Conference on Computer Vision (2010)
39. Wasserman, S., Faust, K., Iacobucci, D.: Social Network Analysis: Methods and Applications. Cambridge University Press, Cambridge (1994)
40. Weng, C.-Y., Chu, W.-T., Wu, J.-L.: Rolenet: Movie analysis from the perspective of social networks. IEEE Trans. Multimedia **11**(2), 256–271 (2009)
41. Yanagawa, A., Chang, S.-F., Kennedy, L., Hsu, W.: Columbia university's baseline detectors for 374 lscom semantic visual concepts. Technical report, Columbia University (2007)
42. Yang, T., Chi, Y., Zhu, S., Gong, Y., Jin, R.: A bayesian approach toward finding communities and their evolutions in dynamic social networks. In: SIAM Conference on Data Mining (2009)
43. Yilmaz, A., Shah, M.: Recognizing human actions in videos acquired by uncalibrated moving cameras. In: International Conference on Computer Visioniccv (2005)
44. Yilmaz, A., Shah, M.: A differential geometric approach to representing the human actions. Comput. Vis. Image Underst. **109**(3), 335–351 (2008)
45. Yu, T., Lim, S.-N., Patwardhan, K., Krahnstoever, N.: Monitoring, recognizing and discovering social networks. In: IEEE Conference on Computer Vision and Pattern Recognition (2009)
46. Zhai, Y., Shah, M.: Video scene segmentation using markov chain monte carlo. IEEE Trans. Multimedia **8**(4), 686–697 (2006)
47. Zhang, D., Gatica-Perez, D., Bengio, S., McCowan, I.: Modeling individual and group actions in meetings with layered hmms. IEEE Trans. Multimedia **8**(3), 509–520 (2006)

Chapter 3
Community Understanding in Location-based Social Networks

Yi-Liang Zhao, Qiang Cheng, Shuicheng Yan, Daqing Zhang
and Tat-Seng Chua

3.1 Background

The past decade has seen a rapid development and change of the Web and Internet
and we are currently witnessing an explosive growth in the social Web, where large
amounts of social media contents are being created constantly by a globally distributed array of disparate sensors including human-sensors, mobile phones, video
cameras, etc. Numerous participatory Web and social networking sites have been
cropping up, empowering new forms of collaboration, communication, and emer-
gent intelligence. Online social networks (OSNs), such as Twitter, Facebook, and
Flickr are reporting millions of new user posts each day.

Specifically, with the high penetration of GPS-enabled smart phones in recent
years, we have witnessed a boom in location-based social networks (LBSNs), such
as Gowalla, Whrrl and Foursquare. In LBSNs, users can perform check-ins,[1] post
comments and upload photos while these pieces of information are immediately dis-
seminated via social graphs to their friends and public. In the context of Foursquare,

---

[1] A check-in is a user's status message in LBSNs with the purpose of letting friends/public
know her current location.

Y.-L. Zhao (✉) · T.-S. Chua
National University of Singapore, Computing 1, 13 Computing Drive,
Singapore 117417, Singapore
e-mail: zhaoyiliang@gmail.com

Q. Cheng
IBM Austrialia, 601 Pacific Highway, St Leonards, NSW 2065, Australia

S. Yan
National University of Singapore, Block E4, #08-27, 4 Engineering Drive 3,
Singapore 117583, Singapore

D. Zhang
Telecom SudParis, Institut Mines-Télécom/Télécom SudPais, 9, rue Charles Fourier,
91011 Evry Cedex, France

Y. Fu (ed.), *Human-Centered Social Media Analytics*,
DOI: 10.1007/978-3-319-05491-9_3, © Springer International Publishing Switzerland 2014

user comments are termed tips, which may cover a variety of diverse topics.[2] Photos, on the other hand, visually present the interesting aspects of the venues visited. Both tips and photos in LBSNs bring together rich user-generated multimedia contents to enrich venue semantics [23] as well as users' profiles.

Community detection is a classical task in social network analysis [26, 38]. With the expanded use of Web and social media, virtual communities and online interactions have become a vital part of human experience and the reflections of people's social positions.[3] Members of same virtual communities tend to share similar interests or topics, and connect to each other more frequently within the community. Given a graph, where vertices represent users and edges represent interactions among users in assorted forms, the goal of community detection is to partition the graph into dense regions of subgraphs, where these dense regions are expected to correspond to users who are closely related, and hence are considered to belong to a community [28]. In OSNs, users may be related explicitly by friendship (Facebook), following/followed (Twitter) or implicitly by common tagging behaviors (Flickr) or similar likes/dislikes (YouTube). Communities of users who are more similar can then be identified by various graph clustering algorithms, where good performances are reported [8].

Effective community detection enables further analysis on community understanding, which helps to figure out the underlying reasons why users are connected and their common characteristics in the same community. Some OSNs, such as Flickr have manually created user communities, where the community profile or the common interests of its members can be deduced by their contributed photos or simply the name of the group. Examples of these communities are animal lovers,[4] nature lovers,[5] etc. In LBSNs, communities are usually location or activity- oriented due to the location-centric nature of the users' interactions and may include shoppers, travelers, food lovers, etc.

The success of community understanding greatly helps social search [1], recommendations [43] and personalization [6]. For example, the work in [43] showed significant improvement on recommendation performance by first identifying meaningful subgroups of user and items. The involvement of community information also shows its potential in more effective targeted advertising [19]. Besides, community information can be utilized to assist and lead to more targeted multimedia annotations [17, 36]. If we know the domain of the multimedia data to be annotated, we can rely more on the inputs from users from the same domain. For example, nature lovers are likely to provide more precise and informative tags for photos showing nature scenes. In addition, more novel multimedia applications, such as mobile collective recommendations [45, 47] and social-aware place visualizations [7, 44] can be enabled by successful and effective community detection in LBSNs.

---

[2] For example, a tip left at an art museum may recommend a special exhibition or give positive/negative comments on the museum environment.

[3] http://en.wikipedia.org/wiki/Social_position

[4] http://www.flickr.com/groups/animalia/

[5] http://www.flickr.com/groups/11611663@N00/

3.2 Motivation

The huge amount of location-tagged multimedia data generated by Foursquare users provides us unprecedented opportunities to understand collective user behaviors on a large scale. The voluminous number of UGCs and fast expansion of network diameter challenge us to perform social multimedia and network analysis [4, 14, 38] on hundreds of thousands to even millions of entities. One fundamental task in social media network analysis is to understand human collective behaviors by identifying people' *social positions*[6] or cohesive subgroups whose group members interact with each other more frequently than those outside the group [8, 9, 22]. Different from OSNs (e.g., LinkedIn, Facebook) which usually have explicit groups for users to join, LBSNs have no explicit community structure.

While previous works have reported promising results on clustering communities from traditional social networks [9, 38], the heterogeneous user behaviors in LBSNs bring together both "virtual" and "physical" interactions, which make it very challenging to develop new frameworks to model the network in a natural and unified manner for community detection. Figure 3.1 shows a snapshot of typical user behaviors in Foursquare, which might correspond to certain overlapping communities we aim to detect.

In addition, most state-of-the-art community detection approaches extract community structures by minimizing certain objective functions while ignoring the equally if not more important task of "understanding" the characteristics of the groups [12]. These approaches find communities based on the intuition that a "good" community should have its members densely connected internally and sparsely connected with other communities [9]. However, as pointed by Fortunato, there is no guarantee that these approaches can provide good quality detection [8]. Guimera et al. also reveal that a maximum modularity may not imply that true community structure is discovered, since random networks may also contain high modularity partitions [10]. Though Tang et al. have attempted to profile the mined communities by extracting descriptive features by using some heuristics [32], it remains unclear what are the underlying reasons to bind the members together and how to interpret the community profiles in terms of the extracted features.

3.3 Challenges

In LBSNs, the heterogeneous interactions among multiple types of entities such as users, tips, venues, and photos naturally form a multimodality and nonuniform hypergraph, where each modality corresponds to one type of entities and each hyperedge connects varying number of entities from different modalities. For example, check-ins in LBSN connect venues to users while tips/photos connect textual topics/visual concepts to users and venues. These kinds of interactions are naturally represented

---

[6] http://en.wikipedia.org/wiki/Social_position

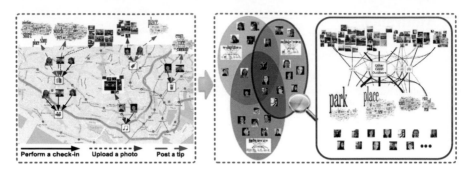

Fig. 3.1 Overview of the profilable and overlapping community detection problem. (*Left*) Heterogeneous users' behaviors in Foursquare. Users can perform check-ins (*black arrows*), post comments (*blue arrows*), and/or upload photos (*red arrows*) at various venues. (*Right*) The detected overlapping communities. Each ellipse represents a profilable community whose characteristics are shown through the tripartite inter-entity relation graph. Each user may belong to one or more communities. (Best view in color)

by hyperedges with nonuniform affinity relations. In addition, there also exist latent relations among venues. For example, grocery stores are more similar to supermarkets than to parks. To utilize the traditional community detection approaches such as modularity optimization [22] or other heuristics-driven methods, such as [5, 9], we need to reduce the complex hypergraph to simpler bipartite or one-modality graph through processes such as "flattening" or "projection" with possible information loss [21, 46]. Thus, we need a flexible mechanism to both simultaneously model these heterogeneous entities and interactions and be able to directly identify and understand communities without any amendments of network structures.

3.4 Overview

To tackle these challenges, we propose a novel and unified framework which performs both community detection and group profiling in LBSNs. We first construct a heterogeneous, multimodality and nonuniform hypergraph which naturally captures various kinds of interactions, such as check-in actions or tip-posting actions in LBSNs. We then propose an efficient algorithm to discover multiple overlapping communities by constraining the minimum number of entities in each modality. The advantages of our proposed method are multifold: (1) The method is general, which can be used in any community detection tasks as long as the network structure is represented as a hypergraph. Moreover, it allows new types of interactions and modalities to be easily added with the technology advancement and emerging of new services; (2) The approach is able to automatically determine the number of interest communities with overlapping entities, and (3) Group profiling can be easily performed since the final computed community contains both users and the "reasons" why they are put in that particular community. In the context of LBSNs, the "reasons" could be the combination of venues they visit, tips they post, and photos they upload.

3.5 Heterogeneous Hypergraph Construction

The heterogeneity of activities in LBSNs naturally brings multiple types of entities and interactions into the same network, which we call a multimodality hypernetwork where each modality corresponds to one type of entity. In Foursquare network, there are four modalities: user, venue, tip, and photo, and three types of interactions: a user checks in at a venue, posts a tip, and uploads a photo at a venue. In addition, entities of same modality may also be related. Figure 3.2 illustrates typical interactions and inter venue category connections in Foursquare. These heterogenous interactions naturally lead to the construction of a nonuniform and multimodality hypergraph. In this section, we first introduce the different types of vertices (Sect. 3.5.1) and hyperedges (Sect. 3.5.2) involved in the hypergraph. We then give details on how we construct each type of hyperedge (Sects. 3.5.3, 3.5.4, 3.5.5 and 3.5.6).

3.5.1 Vertices

There are four types of vertices involved in interactions in LBSNs: user, venue, tip, and photo, as is shown in Fig. 3.2. Formally, let $\mathbb{V}$ be the vertex set, which can be divided into g subsets, i.e., $\mathbb{V} = \bigcup_{a=1}^{g} \mathbb{V}_a$. In Foursquare network, $g = 4$ and each subset $\mathbb{V}_a$ corresponds to set of vertices of modalities: user, venue, tip, and photo, respectively. Let $\mathbb{V}_1 = \mathbb{U} = \{u_1, \ldots, u_{n_u}\}$, $\mathbb{V}_2 = \mathbb{L} = \{l_1, \ldots, l_{n_l}\}$, $\mathbb{V}_3 = \mathbb{T} = \{t_1, \ldots, t_{n_t}\}$, and $\mathbb{V}_4 = \mathbb{D} = \{d_1, \ldots, d_{n_d}\}$ be sets of users, venues, tips and photos respectively, where n_a is the number of vertices in $\mathbb{V}_a$, $a \in \{1, 2, 3, 4\}$.

3.5.2 Hyperedges

There are three types of interactions involved in the Foursquare network: a user checks in at a venue, posts a tip, and uploads a photo at a venue. Thus, the first three types of hyperedges correspond to these three types of interactions. Formally, let $\mathbb{E}$ be the hyperedge set, which can be divided into s subsets, i.e. , $\mathbb{E} = \bigcup_{b=1}^{s} \mathbb{E}_b$, with each hyperedge representing a n_b-ary affinity relation. In Foursquare network, $s = 4$. We build three sets of hyperedges corresponding to the three types of interactions: $\mathbb{E}_1 = \{(u_i, l_j)\}$ representing a check-in performed by user u_i at venue l_j, $\mathbb{E}_2 = \{(u_i, t_j, l_k)\}$ representing tip t_j is posted by user u_i at venue l_k and $\mathbb{E}_3 = \{(u_i, d_j, l_k)\}$ representing photo d_j is uploaded by user u_i at venue l_k. In addition, we also want to

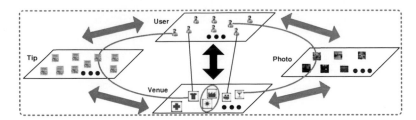

Fig. 3.2 Heterogeneous entities and interactions in Foursquare. The *red line* and *arrows* represent interactions among users, photos, and venues; The *blue line* and *arrows* represent interactions among users, tips, and venues; The *black line* and *arrows* represent interactions between users and venues; And the *green circle* connects venues with similar functions. *Note* For color interpretation see online version

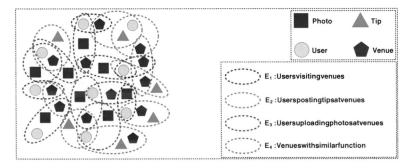

Fig. 3.3 The illustration of hypergraph construction. Vertices of *circle*, *triangle*, *square*, and *pentagon* represent entities of modality: user, tip, photo, and venue, respectively. Hyperedges represented by *ellipses* of *red*, *green*, *blue* and *tan* represent interactions of venue visiting, tip posting, photo uploading and similar venues, respectively

model inter-venue similarities and define hyperedge set $\mathbb{E}_4 = \{(l_1, \ldots, l_h)\}$. We can then denote the hypergraph as $\mathbb{G} = \{\mathbb{V}, \mathbb{E}, w\}$, where $w : \mathbb{E} \to \mathbb{R}^+$ is a weighting function which associates a positive real value with each hyperedge, with larger weights representing stronger affinity relations. Figure 3.3 illustrates hypergraph constructions for Foursquare hypernetwork.

Given the enormous number of entities in each modality, the hypergraph is extremely sparse, which unavoidably weakens the structure information. Thus, to better characterize different types of interactions, we seek to first group semantically similar entities together in each modality to construct a denser hypergraph. The following subsections detail the constructions of each type of hyperedge.

3.5.3 User-Venue Hyperedges: $\mathbb{E}_1$

To group similar venues, we refer to their corresponding venue categories. Foursquare maintains nine predefined root categories[7] and a hierarchical structure of venue categories with a total of 435 venue categories including the root categories as of now. We consider two venues to be the same if they belong to the same leaf venue category. For simplicity, we use l to represent venue category in the rest of the chapter. Thus, each edge $(u_i, l_j) \in \mathbb{E}_1$ indicates a check-in performed by user u_i at venue category l_j and $w((u_i, l_j)) = c(u_i, l_j)$, where $c(u_i, l_j)$ is the number of check-ins logged by u_i at l_j.

3.5.4 User-Tip-Venue Hyperedges: $\mathbb{E}_2$

The voluminous number of tips make it difficult to model correlations among users without modeling tips' similarity explicitly. We seek to first extract a middle-level representation of tips and then directly relate users to the extracted representations. In this way, we are able to reduce the number of noisy hyperedges significantly and obtain better interpretations of heterogenous comments posted by users at various venues. To do so, we first project each tip to a latent topic space using latent dirichlet allocation (LDA), which is able to mine higher level representations, named "topics," from a collection of documents [2]. Essentially, LDA helps to explain the similarity of tips by grouping tips into latent sets (topics). A mixture of these sets then constitute the observed tips.

We use MALLET [18] to train the topic model with 100 topics for 2,000 iterations, optimize the parameters every 50 iterations, and update the user-tip-venue hyperedges by replacing each tip with its corresponding topics. Formally, if a tip t is posted by user u at venue l, and t is mapped to a series of topics z_m, $m = 1, ...,100$, then we construct a series of user-tip-venue hyperedges as (u, z_m, l), $m = 1, ...,100$, with weight set as $w((u, z_m, l)) = \sum_t p(z_m|t, u, l)$, where $p(z_m|t, u, l)$ is the topic distribution of tip t posted by user u and venue l. For simplicity, we use t to represent tip topic in the rest of the chapter.

To validate the interpretability of the extracted latent topics, we first generate a word cloud for each topic using Wordle.[8] For each topic t, the more prominent words have larger font size. We then compute the conditional probabilities of venue categories given each topic to investigate the semantic correlations between topics and venue categories as follows. Given a tip topic t, its correlation with venue category l is computed as: $p(l|t) = \frac{\sum_{u \in \mathbb{U}} w((u,t,l))}{\sum_{l' \in \mathbb{V}} \sum_{u \in \mathbb{U}} w((u,t,l'))}$. Figure 3.4a shows three extracted topics along with the correlation visualization of the venue categories. Venue categories with larger size and shorter/thicker edges are those that are more related

---

[7] http://aboutfoursquare.com/foursquare-categories/

[8] http://www.wordle.net

with the corresponding topics. We observe that people do discuss topics that are semantically related to the venues they visit. For example, the venues where people discuss night-life related topics are bars, night clubs, etc. And people usually discuss movie-related topics at movie theaters and multiplexes.

3.5.5 User-Photo-Venue Hyperedges: $\mathbb{E}_3$

The "User-Photo-Venue" edges aim to express what kinds of photos users captured at different venues. Unlike previous section where we project each tip to a latent topic space, we explicitly map each photo to a predefined concept list. The concept list is constructed through supervised learning on a large dataset comprising many labels. The reasons to use a predefined concept list rather that latent mining are as follows. First, the image classification task has achieved great improvement and many concepts of images can be well-trained and generalized. Second, the result of latent mining of image concept is hard to depict and does not have obvious semantic meaning.

We consider two kinds of concepts: scenes and objects. Most of the images in Foursquare are taken at venues and with related objects, so it is rather important to determine what kinds of scenes/objects are contained in the images. For scene categories, we select a list of scene categories from SUN Scene dataset [39]. The dataset contains various indoor and outdoor scenes of 899 categories and 130, 519 images, such as balcony, beach, bridge, building, park, street, and so on. These categories are quite consistent with the venue categories in the Foursquare dataset. We select 63 categories of the SUN dataset which have exact or similar names with the venue categories and obtain a subset with a total of 39, 782 images. For objects, we select the most practical object recognition dataset in computer vision area, i.e., the PASCAL VOC *2010* dataset [33]. This dataset aims to recognize objects from a number of visual object classes in realistic scenes (i.e., not presegmented objects) and provides a training set of 21, 738 images. The dataset has four main categories, including person, animal (bird, cat, etc.), vehicle (aeroplane, bicycle, train, etc.), and indoor (bottle, chair, sofa, tv/monitor, etc.). We make use of all 20 object classes from these four categories. Thus, there are totally $63 + 20 = 83$ concept categories in our training set.

We train the 83 concept categories in a supervised manner. For each image, we first extract dense SIFT descriptors. The implementation of dense SIFT is based on VL-Feat [35] using multiple scales setting (spatial bins are set as 4 and 8) with step size of 4. We use the improved Fisher vector coding [27] which has demonstrated the superiority over other coding methods in a fair setting [11]. The component number of Gaussian mixture model in Fisher vector coding is set to 128. One-versus-All SVM is learnt for each category in SUN Scene dataset and PASCAL VOC *2010* dataset. We also perform 10-fold cross validation to get the classification response over the training dataset. Then we learn a probability mapping from the classification output score s. For each concept category c_i, we first get a threshold θ_i with maximum

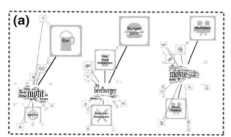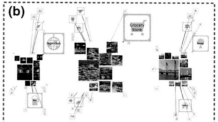

Fig. 3.4 a Night Life (*left*), Western Food (*middle*), Movie (*right*). **b** Night Club (*left*), Shop (*middle*), Bridge (*right*). Three of the tip topics/photo concepts and their correlations with the top 20 venue categories. For each tip topic/photo concept, venue categories with stronger correlations with the topics/concepts are presented with larger size and with shorter/thicker connections. (Best view in high 200 % resolution in Acrobat Reader)

F1 score measurement on the training set, then the probability of photo d containing concept c_i is defined as: $p(c_i|d) = \frac{1}{1+e^{-\gamma s_i}}$, if the classification output score $s_i \geq \theta_i$ and 0 otherwise.

Once we obtain the concept probability of each photo, we update the user-photo-venue hyperedges by replacing each photo with its corresponding concepts. Formally, if photo d is uploaded by user u at venue l and d is mapped to a series of concepts $c_m, m = 1, \ldots, M$, then we construct a series of user-photo-venue hyperedges as $(u, c_m, l), m = 1, \ldots, M$ with each weight computed as $w((u, c_m, l)) = \sum_d p(c_m|d, u, l)$, where $p(c_m|d, u, l)$ is the concept distribution of photo d uploaded by user u and venue l. For simplicity, we use d to represent photo concept in the rest of the chapter.

To visualize the semantic relatedness between photo concepts and venue categories, we first generate a photo concept cloud analogous to the topic word cloud for each concept category. For each photo concept, the more prominent photos have larger size. We then compute the conditional probabilities of venue categories given each photo concept as follows. Given an photo concept d, its correlation with venue category l is computed as: $p(l|d) = \frac{\sum_{u \in \mathbb{U}} w((u,d,l))}{\sum_{l' \in \mathbb{V}} \sum_{u \in \mathbb{U}} w((u,d,l'))}$. Figure 3.4b shows three photo concepts along with the correlation visualization of the venue categories. Venue categories with larger size and shorter/thicker edges are those that are more related to the corresponding concepts. We observe that the photos are generally semantically related to the venue categories where the photos are taken.

3.5.6 Venue-Venue Hyperedges: $\mathbb{E}_4$

In addition, we also consider inter-category similarity by creating hyperedges among venue categories with similar functions. For example, "Concert Halls" are more similar to "Jazz Clubs" than to "Baseball Stadiums". Without loss of generality, we create hyperedges among venue categories of the same parent in the second level of

Foursquare category hierarchy. Thus, after mapping each venue to its category, we define $\mathbb{E}_4 = \{(l_1, \ldots, l_h)\}$ where $l_1 \cdots l_h$ share the same parent category in the second level of Foursquare category hierarchy. We initialize the weights of all hyperedges in $\mathbb{E}_4$ to be 1.

3.6 Community Detection Over Heterogeneous Graph

In this section, we formulate the profilable and overlapping community detection task as a problem of dense subgraph detection over heterogeneous hypergraph (Sect. 3.6.1), and develop an efficient algorithm to solve this optimization problem (Sect. 3.6.2) motivated by [15].

3.6.1 Problem Formulation

The constructed interactions involving various types of entities from different modalities are modeled by a nonuniform heterogenous hypergraph $\mathbb{G} = \{\mathbb{V}, \mathbb{E}, w\}$. As is defined in Sect. 3.5.1, $\mathbb{V} = \bigcup_{a=1}^{g} \mathbb{V}_a$ is a finite set of vertices, where each $\mathbb{V}_a$ is a subset of vertices from different modalities. In this problem, $g = 4$ and $\mathbb{V}_1, \mathbb{V}_2,$ $\mathbb{V}_3$ and $\mathbb{V}_4$ correspond to modality of user, venue, tip, and photo, respectively. Let $n = n_u + n_l + n_t + n_d$ be the total number of vertices, where $n_u, n_l, n_t,$ and n_d are the numbers of vertices of modality: user, venue, tip, and photo, respectively as is defined in Sect. 3.5.1. Let $\mathbb{E} \subseteq \mathbb{V}^{n_e}$ be the set of all hyperedges, with each hyperedge $e \in \mathbb{E}$ relating to n_e entities and representing an n_e-ary affinity relation. $\mathbb{V}^{n_e}$ represents the set of vertices involved in the n_e-ary affinity relation. Following [15], we define the density of a subgraph as:

$$f(\mathbf{x}) = \sum_{s=1}^{|\mathbb{E}|} w_s \overbrace{x_{s_1} \ldots x_{s_{n_s}}}^{n_s}, \qquad (3.1)$$

where w_s is the weight of hyperedge s involving vertices: $(x_{s_1}, \ldots, x_{s_{n_s}})$, n_s is the number of vertices involved in hyperedge s and $\mathbf{x}$ is a vector with n components, with each x_i representing the probability of choosing the ith vertex of $\mathbb{V}$ and $\sum_i x_i = 1$.

We formulate the profilable and overlapping community detection problem as a dense subgraph detection problem, where each dense subgraph of the hypergraph defines one community and its profiles. Inspired by [15], our aim is to find a subgraph consisting of vertices participating in the densest interactions within a local region. Suppose a subset $\mathbb{C} \subseteq \mathbb{V}$ includes vertices which form a dense subgraph and $\mathbb{E}_{\mathbb{C}}$ is the corresponding edge set. If $\mathbb{C}$ is really a dense subgraph, then most of hyperedges in $\mathbb{E}_{\mathbb{C}}$ should have large weights, which implies that $f(\mathbf{x})$ should be relatively large. Thus, the dense subgraph detection problem corresponds to the problem of maximizing $f(\mathbf{x})$ with the natural constraints as follows.

$$\max f(\mathbf{x}) = \sum_{s=1}^{|\mathbb{E}|} w_s \overbrace{x_{s_1} \dots x_{s_{n_s}}}^{n_s} \tag{3.2a}$$

$$\text{s.t.} \sum_{i \in V} x_i = 1, \tag{3.2b}$$

$$x_i \in [0, 1]. \tag{3.2c}$$

The above formulation is able to find dense and overlapping subgraphs given multiple initializations. However, our aim is not only to find dense and overlapping communities but also to make the detected communities profilable, which means that it is easy to understand and interpret what are the common characteristics and human behaviors in each community. Information that helps to profile communities comes from modalities showing people's interests. For example, members of a night life club tend to visit bars, night clubs, pubs, etc. In addition, they tend to take photos and discuss topics related to night life. To help better understand the detected communities, we want to include sufficient number of entities from various modalities other than users in the detected communities. Moreover, we would like to control the community size. Depending on the applications, users may be interested in communities detected at different scales. For example, sometimes, we would like to find communities which reflect human behaviors at city level. In this situation, it only makes sense if the number of members is significantly large and we do not want to find communities which only involve a couple or a family where intensive interactions take place on a daily basis. As a result, it is desirable to make the community size controllable. These two requirements motivate us to add two more constraints to the optimization problem:

$$\max f(\mathbf{x}) = \sum_{s=1}^{|\mathbb{E}|} w_s \overbrace{x_{s_1} \dots x_{s_{n_s}}}^{n_s} \tag{3.3a}$$

$$\text{s.t.} \sum_{i \in V} x_i = 1, \tag{3.3b}$$

$$\sum_{j \in V_k} x_j \geq c_k, \tag{3.3c}$$

$$x_i \in [0, \varepsilon]. \tag{3.3d}$$

In (3.3c), V_k is the vertex set of modality k, where $k \in \{1, 2, 3, 4\}$. The objective function (3.3a) prefers vertices connecting to many hyperedges with large weights. The constraint (3.3b) together with the constraint (3.3d) requires that each detected community contains at least $\lceil \frac{1}{\varepsilon} \rceil$ vertices. It is worth noting that we are dealing with heterogeneous hypergraph and it is possible that some entities from modality k have little interaction with other entities in some subgroups. As mentioned above, to prevent that the final subgraph has few/none vertices of certain modality, which makes it difficult to profile communities, we set a constraint for each modality, i.e.,

$\sum_{j \in \mathbb{V}_k} x_j \geq c_k$ where c_k is the lower bound of the existence probability of modality k in the final detected communities. Thus, there is at least $\lceil \frac{c_k}{\varepsilon} \rceil$ vertices of modality k in the detected communities. Obviously, $\sum_{k=1}^{g} c_k \leq 1$, otherwise the optimization problem (3.3) will have no solution.

Each community corresponds to one local maximum of (3.3). To obtain all important communities, we need to find all significant local maxima of (3.3). We adopt a similar approach as in [15], that is, systematically generating many initializations, and then efficiently approach a local maximum from each initialization. In this chapter, we construct the initialization $\mathbf{x}(0)$ as follows. We first build a bipartite graph by considering only interactions between users and venues, i.e., users performing check-ins. We then project the constructed bipartite graph to a one-modality graph consisting of only users and use Tang and Liu's edge clustering approach to initialize a list of K overlapping initial user groups [30]. We then assign entities from venue categories, tips, and photos to each of the initial group based on the users who are involved in the interactions according to the initialization.

The final obtained points $\mathbf{x}^*$ are usually local maximizers of (3.3), and thus good candidates of underlying communities. Unlike [15], which works on the uniform and homogeneous hypergraphs, here we make two extensions: (1) the constructed hypergraph is heterogeneous and nonuniform, and (2) we explore new constraints, namely, we require that a minimum number of entities of each modality should co-exist in the final dense subgraphs.

3.6.2 Optimization

Since the problem (3.3) is a constrained optimization problem, by adding Lagrangian multipliers λ, $\alpha_i \geq 0$ and $\beta_i \geq 0$ for all $i = 1, \ldots, n$, π_i for all $i = 1, \ldots, g$ ($g = 4$ in this problem), we obtain its Lagrangian function:

$$L(\mathbf{x}, \lambda, \boldsymbol{\alpha}, \boldsymbol{\beta}, \boldsymbol{\pi}) = f(\mathbf{x}) - \lambda \left(\sum_{i=1}^{n} x_i - 1 \right) + \sum_{i=1}^{n} \alpha_i x_i + \sum_{i=1}^{n} \beta_i (\varepsilon - x_i)$$
$$+ \sum_{k=1}^{g} \pi_k \left(\sum_{j \in \mathbb{V}_k} x_j - c_k \right). \tag{3.4}$$

Any local maximizer $\mathbf{x}^*$ must satisfy the Karush-Kuhn-Tucker (KKT) condition, i.e., the first-order necessary conditions for local optimality. That is,

$$\begin{cases} g_i(\mathbf{x}^*) - \lambda + \alpha_i - \beta_i + \pi_k = 0, i = 1, \ldots, n, \ v_i \in \mathbb{V}_k \\ \sum_{i=1}^{n} x_i^* \alpha_i = 0 \\ \sum_{i=1}^{n} (\varepsilon - x_i^*) \beta_i = 0 \\ \sum_{k=1}^{g} (\sum_{j \in \mathbb{V}_k} x_j^* - c_k) \pi_k = 0, \end{cases} \tag{3.5}$$

where v_i, $i \in \{1, \ldots, n\}$ is a vertex with the corresponding probability: x_i and $g_i(\mathbf{x}) = \frac{\partial f(\mathbf{x})}{\partial x_i}$. Since x_i, α_i, β_i and π_k are all nonnegative, $\sum_{i=1}^{n} x_i^* \alpha_i = 0$ is equivalent to saying that if $x_i^* > 0$, then $\alpha_i = 0$, and $\sum_{i=1}^{n} (\varepsilon - x_i^*) \beta_i = 0$ is equivalent to saying that if $x_i^* < \varepsilon$, then $\beta_i = 0$, and $\sum_{k=1}^{g} (\sum_{j \in \mathbb{V}_k} x_j^* - c_k) \pi_k = 0$ is equivalent to saying that if $\sum_{j \in \mathbb{V}_k} x_j^* > c_k$, then $\pi_k = 0$. Hence, the KKT conditions can be rewritten as:

$$g_i(x^*) \begin{cases} \leq \lambda - \pi_k, & x_i^* = 0, v_i \in \mathbb{V}_k; \\ = \lambda - \pi_k, & x_i^* \in (0, \varepsilon), v_i \in \mathbb{V}_k; \text{ and } \pi_k \begin{cases} = 0, & \sum_{j \in \mathbb{V}_k} x_j^* > c_k; \\ \geq 0, & \sum_{j \in \mathbb{V}_k} x_j^* = c_k. \end{cases} \\ \geq \lambda - \pi_k, & x_i^* = \varepsilon, v_i \in \mathbb{V}_k; \end{cases}$$

(3.6)

As pointed out by [15], we can optimize the problem (3.3) in the following pairwise way. That is, each time we only update a pair of components (x_i, x_j):

$$x_r^{\text{new}} = \begin{cases} x_r, & r \neq i \text{ and } r \neq j; \\ x_r + \mu, & r = i; \\ x_r - \mu, & r = j. \end{cases}$$

(3.7)

After updating x by (3.7), the change in value of function $f(\mathbf{x})$ is:

$$f(\mathbf{x}^{\text{new}}) - f(\mathbf{x}) = -g_{ij}(\mathbf{x}) \mu^2 + (g_i(\mathbf{x}) - g_j(\mathbf{x})) \mu,$$

(3.8)

where $g_{ij}(\mathbf{x}) = \frac{\partial^2 f(\mathbf{x})}{\partial x_i \partial x_j}$ and $g_j(\mathbf{x}) = \frac{\partial f(\mathbf{x})}{\partial x_j}$.

When $g_i(\mathbf{x}) > g_j(\mathbf{x})$, we may increase $f(\mathbf{x})$ by (3.7). However, μ is also affected by the constraints of (3.3). The constraint (3.3b) is always satisfied. To satisfy other constraints, there are two situations: (1) if v_i and v_j belong to the same modality, then the constraint (3.3c) is satisfied, we can set $\mu = \min \left\{ x_j, \varepsilon - x_i, \frac{g_i(\mathbf{x}) - g_j(\mathbf{x})}{2 g_{ij}(\mathbf{x})} \right\}$ to maximize the increase of the objective function (3.3). (2) if v_i and v_j belong to different modalities, then we need to satisfy the constraint (3.3c), thus we set $\mu = \min \left\{ x_j, \varepsilon - x_i, \frac{g_i(\mathbf{x}) - g_j(\mathbf{x})}{2 g_{ij}(\mathbf{x})}, \sum_{v_l \in \mathbb{V}_k, v_j \in \mathbb{V}_k} x_l - c_k \right\}$. We then define the set $\mathbb{U}$ as: $\mathbb{U} = \left\{ (v_i, v_j) \big| g_i(\mathbf{x}) > g_j(\mathbf{x}), x_i < \varepsilon, x_j > 0, \sum_{v_l \in \mathbb{V}_k} x_l > c_k, \text{ if } v_i \notin \mathbb{V}_k, v_j \notin \mathbb{V}_k \right\}$.

Obviously, $\mathbb{U}$ is the set of pairs (v_i, v_j) which can increase $f(\mathbf{x})$ by (3.7). The theorem below establishes the relation between the KKT point x of (3.3) and the set $\mathbb{U}$, which is the basis of our optimization method.

Theorem 1 $\mathbf{x}$ *is a KKT point of* (3.3) *iff* $\mathbb{U} = \emptyset$.

The proof of this theorem is obvious according to the KKT condition (3.6), thus we omit it here. According to Theorem 1, from any initialization $\mathbf{x}(0)$, we can iteratively choose a pair from $\mathbb{U}$ and optimize (3.3) according to (3.8). This process terminates until the set $\mathbb{U}$ is empty, that is, a KKT point has been reached. Algorithm 1 summarizes the whole procedure. Intuitively speaking, Algorithm 1 successively chooses the "good" vertex and the "bad" vertex and then updates their corresponding components of $\mathbf{x}$, that is, increases the probability of choosing the "good" vertex and decreases the probability of choosing the "bad" vertex. Algorithm 1 is highly efficient

---
Algorithm 1: Compute a KKT point x from an initialization $\mathbf{x}(0)$

1: **Input:** The optimization problem (3) and an initialization $\mathbf{x}(0)$;
2: Set $\mathbf{x} = \mathbf{x}(0)$;
3: **repeat**
4:     Update the partial derivative $g_i(\mathbf{x})$ with respect to each variable x_i;
5:     Find a pair $(v_i, v_j) \in U$ and compute the best step size μ to maximize the increase the objective function (3);
6: **until** $\mathbf{x}$ is a KKT point
7: **Output:** A KKT point $\mathbf{x}$.

---

since we only work on a small dense subgraph in each iteration. Only two components of $\mathbf{x}$ are changed, thus only the partial derivatives of a small set of components of $\mathbf{x}$ are affected. Moreover, the proposed procedure can be easily implemented in parallel when there are huge number of initializations.

From each initialization, we can obtain a local KKT point of (3.3), which usually represents a community. Since we optimize (3.3) from many initializations, we obtain many communities. Note that some communities may be duplicate and we need to eliminate the duplications. Some communities may overlap, which is in fact the advantage of our method, since real communities may overlap. Since $f(\mathbf{x})$ measures the degree of connectedness in each dense subgraph, it is a natural measure to rank all communities. And the larger the function value $f(\mathbf{x})$ is, the higher the probability of $\mathbf{x}$ represents a real community.

3.7 Empirical Evaluation

In this section, we first introduce the dataset and experimental setup in Sect. 3.7.1. We then conduct a comprehensive set of experiments to evaluate our proposed approach on three tasks: (1) prediction of users' visits, (2) photos' concept annotation, and (3) prediction of what users discuss at various venues in Sect. 3.7.2. Finally, we present the visualization of the detected social communities at the global scale and two city scales in Sect. 3.7.3.

3.7.1 Dataset and Experimental Setups

Since Foursquare API provides limited authorized access for retrieving check-in information, we resort to Twitter streaming API[9] to get the publicly shared check-ins. We have recorded more than 6 million check-ins generated by more than one

---
[9] https://dev.twitter.com/docs/streaming-api

Table 3.1 Foursquare dataset for the profilable and overlapping community detection task

	Users	Check-ins	Tips	Images
Global	13,068	86,302	335,877	69,510
Singapore	8,736	32,156	156,761	9,775
New York city	9,918	51,043	213,302	22,135

hundred thousand users, between January *2012* and March *2012* through the stream of Twitter messages. We then filter out "spam" users, whose average consecutive check-in intervals are less than 1 min. Afterwards, we use Foursquare APIs[10] to retrieve the tips and photos contributed by the remaining users. We compute the activeness score for user u as: activeness$(u) = \alpha\#Check - ins + \beta\#Tips + \gamma\#Photos$. # means the number of user u's check-ins/tips/photos. We set $\alpha = \beta = \gamma = \frac{1}{3}$. We select the top 80 % of users ranked by the users' activeness scores at the global scale and two city scales, respectively. Table 3.1 summarizes the dataset used for the profilable and overlapping community detection task.

We define the initial weight of each hyperedge in $\mathbb{E}_1$, $\mathbb{E}_2$ and $\mathbb{E}_3$ based on the frequencies of the interactions and that of $\mathbb{E}_4$ to be unit weight as in Sect. 3.5. We then normalize each type of hyperedge as follows. We normalize hyperedges in $\mathbb{E}_1$ for each user such that the sum of the weights of all the hyperedges of a user is equal to 1. Similarly, we normalize edges of $\mathbb{E}_2$ and $\mathbb{E}_3$ for each (*user, venue*) pair. Finally, we normalize hyperedges in $\mathbb{E}_4$ by the number of vertices in each edge, such that the weights of hyperedges are inversely proportional to the number of their vertices.

Next, we describe the parameter settings as follows.

- Number of initializations: K. This number is not critical, since our proposed dense subgraph mining algorithm works on a small part of hypergraph corresponding to each initialization and is able to detect overlapping communities. If we set K to be larger, the communities with highest density will not change much. After all, only the top few communities have clear profiles. Thus, we empirically set $K = 200$ in the experiments for both global and city scales.
- Variable to control community size: ε. We set $\varepsilon = \frac{1}{K}\sum_{h=1}^{K}(\frac{1}{\sum_k \#M_{hk}})$, where K is the number of communities in the initialization and $\#M_{hk}$ is the number of entities of modality k in community h in the initialization.
- Variable to control minimum number of modalities in each community: c_k. We set $c_k = \frac{1}{K}\sum_{h=1}^{K}(\frac{\#M_{hk}}{\sum_{k'}\#M_{hk'}})$.

It is worth mentioning that our C++ implementation of the dense subgraph detection is highly efficient. The 200 initialization converges to local KKT points within 10 min in a non-parallel mode on a Intel 3.0 GHZ machine with 4 GB memory.

---

[10] https://developer.foursquare.com/docs/

3.7.2 Quantitative Indirect Evaluation

Since the real-world data we use does not have the ground truth[11] available, we resort to indirectly evaluate our proposed approach by using the discovered social communities to predict users' visits, tips, and photos.

Users' behaviors have strong intercorrelations. Intuitively, users visiting similar venues tend to share similar interests, which are reflected through the topics they discuss and photos they take. For example, we expect shoppers to check in at shopping centers or malls and to discuss shopping-related topics more frequently than other users do. Similarly, animal lovers should often visit parks, zoos with most of their photos containing contents related to nature or animals.

The detected communities should intuitively group users with similar interests together, which makes it interesting and possible to investigate whether the community's profile can help to infer individuals' profiles, such as the venues they visit, the comments they post and the photos they take.

Here we propose to evaluate the community detection performance through three tasks. Given that user u belongs to community $\mathbb{C}$, we aim to predict: (1) what is the most likely venue l that u is going to visit: $p(l|\mathbb{C}, u)$; (2) what kind of photos d that u is most likely to take at venue l: $p(d|\mathbb{C}, u, l)$; and (3) what kind of topic t that u is most likely to discuss at venue l: $p(t|\mathbb{C}, u, l)$.

We term $p(l|\mathbb{C}, u)$, $p(d|\mathbb{C}, u, l)$ and $p(t|\mathbb{C}, u, l)$ the *preferences* of user u. Similarly, $p(l|\mathbb{C})$, $p(d|\mathbb{C})$, $p(t|\mathbb{C})$ are the *preferences* of community $\mathbb{C}$. Let ω be either venue l, tip t or photo d and u^p be u's partial information. For example, u^p could be the subsets of types of photos d that u usually takes and venues l that u usually visits. For $p(d|C, u, l)$ and $p(t|C, u, l)$, u^p includes the specific venue information l. User u's preference $p(\omega|\mathbb{C}, u)$ in community $\mathbb{C}$ can be estimated by the community's preference $p(\omega|\mathbb{C})$ and u's partial information by:

$$p(\omega|\mathbb{C}, u) \doteq p(\omega|\mathbb{C}) + p(\omega|\mathbb{C}, u^p), \qquad (3.9)$$

where $p(\omega|\mathbb{C})$ is the community preference, $p(\omega|\mathbb{C}, u^p)$ is u's preference within the community $\mathbb{C}$ with user's partial information u^p. $p(\omega|\mathbb{C})$ can be obtained by computing the modality probability within community while $p(\omega|\mathbb{C}, u^p)$ is statistically estimated based on u^p within $\mathbb{C}$.

3.7.2.1 Data Preparation

In order to conduct the experiments, we preprocess the raw dataset to obtain a ground-truth dataset as follows. We randomly divide the whole Foursquare dataset into two parts for each task, i.e., the testing set containing x % of task-related information and

---

[11] There is no explicit groups defined in Foursquare.

the remaining data constitute the training set. We perform community detection on the training set and predict the missing information based on (3.9). Here, we consider $x \in \{10, 20, 30, 40, 50\}$.

3.7.2.2 Evaluation Metric

We treat each prediction task as a multilabel classification problem and use the mean average precision (MAP) as the evaluation metric. For each task, given a testing set $\mathbb{T}$, we generate a ranking list of predictions for each item $t \in \mathbb{T}$. Average precision (AP) is obtained for each type of venue categories/tip topics/photo concepts. MAP is the average of APs over the total items in the testing set for each prediction task.

3.7.2.3 Baselines

As mentioned in Sect. 3.8, state-of-the-art approaches are not able to directly handle heterogenous nonuniform hypergraph. Thus, we need to first convert the graph into simpler network types, which can then be used by other community detection techniques. Besides comparing with pairwise settings, we also compare with overlapping (Edge clustering [30]) and nonoverlapping (Modularity Maximization [22]) community detection approaches. Besides comparing with other state-of-the-art approaches, we are also interested in studying the importance of using complete information and the informativeness of different modalities. Thus, we also compare the prediction performances between using different partial information and using the complete information. Specifically, we compare our proposed approach with the following baselines.

- *Hyperedge without tip (HWT)*: We remove all hyperedges related to tip postings and use the remaining information to predict users' visits/tips/photos.
- *Hyperedge without photo (HWP)*: We remove all hyperedges related to photo uploadings and use the remaining information to predict users' visits/tips/photos.
- *Hyperedge without check-in (HWC)*: We remove all hyperedges related to check-ins and use the remaining information to predict users' visits/tips/photos.
- *Pairwise (PW)*: To validate the advantages of using hyperedge model, we compare the prediction performances with a model involving only pairwise edges. To obtain pairwise edges from hyperedges, we follow Neubauer and Obermayer's approach [21]: For each hyperedge (e_i, e_j, e_k), we introduce three edges (e_i, e_j), (e_i, e_k) and (e_j, e_k) where the original edge weight is inherited by the three new pairwise edges.
- *Edge clustering (EC)*: We compare the prediction performances with the initialized overlapping groups which are generated by edge clustering [30].
- *Modularity maximization (MM)*: We compare the prediction performance with modularity maximization [22], where we first convert the heterogeneous nonuniform hypergraph into a one-modality user pairwise graph as follows. We first build

a bipartite graph by considering only interactions between users and venues. We then project the constructed bipartite graph to a one-modality graph consisting of only users and use modularity maximization to form K nonoverlapping communities. We then assign entities from venue categories, tips, and photos to each of the initial group based on the acting users who are involved in the interactions according to the initialization.

3.7.2.4 Performance Comparisons

Figure 3.5 shows the performance of using different methods in the three prediction tasks. We have the following observations.

First, we analyze the impact of using complete and partial information on the performance of the system. (1) Overall, the use of hyperedge with complete information achieves the best performance in all the three tasks for different % of training and testing data. (2) Check-ins carry more information then tip postings and photo uploadings. The performance of predicting users' photos and tips degrade the most when we exclude the hyperedges of type $(user, venue)$. The reason could be that venue categories are keys in our task to connect the other two modalities (tips and photos) as well as profile for each detected communities. (3) In addition, we have observed that hyperedges of type $(user, venue, tip)$ carry more information than those of type $(user, venue, photo)$, which is partly caused by the more number of $(user, venue, tip)$ hyperedges.

Next, we compare the performance of our proposed approach with complete information against the three state-of-the-art approaches. (1) We find that using pairwise graph with complete information is the next most competitive approach. It performs only slightly worse than hypergraph, which is consistent with what Neubauer et al. and Zhou et al. pointed out in [21, 46]. (2) Edge clustering does not perform well since we only use check-in information to group users as an initialization. (3) Modularity maximization performs slightly worse than edge clustering, which shows that overlapping communities better capture users' preferences.

To summarize, the evaluations on three tasks validate both the importance of using hyperedges as well as the effectiveness of our proposed approach in detecting meaningful communities.

3.7.3 Qualitative Community Visualization

In this section, we describe how to visualize the detected social communities in terms of their profiles, which comprises two steps: (1) Representative communities extraction (Sect. 3.7.3.1) and (2) Community Profiling (Sect. 3.7.3.2). In Sect. 3.7.3.3, we then visualize some notable communities at the global scale as well as compare some culture differences between Singapore and New York City by analyzing the top-detected communities.

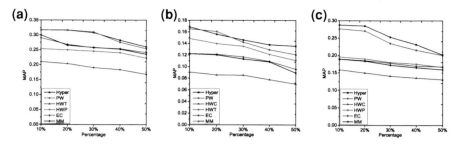

Fig. 3.5 **a** Prediction of users' visits. **b** Prediction of users' photos. **c** Prediction of users' tips. Performance comparisons

3.7.3.1 Representative Communities Extraction

Intuitively, representative communities correspond to the densest subgraphs mined from the reconstructed heterogenous hypergraph with minimum intercommunity overlap. We keep two lists in the extraction process: a candidate list and a selection list, where both lists are complementary to each other. We first add the community with the highest objective function value into the selection list. Then for each remaining communities, we compute their overlapping level with each of the selected communities until the number of selected communities reaches a predetermined value. Without loss of generality, we use Jaccard index to calibrate the overlapping level between two communities: $J(\mathbb{C}_i, \mathbb{C}_j) = \frac{\mathbb{C}_i \cap \mathbb{C}_j}{\mathbb{C}_i \cup \mathbb{C}_j}$.

After we obtain a list of representative communities, we then re-rank the communities based on the number of members in each community. The top ten communities at the global scale and top five communities in Singapore and New York City are presented in the online appendix.

3.7.3.2 Community Profiling

Community profiles are characterized by the properties and inter-relation of the community's dominant nonuser entities from each modality, i.e., venues, tips, and photos. To profile and visualize each representative community, we first compute the importance scores of all nonuser entities from each modality and then visualize each community by constructing a tripartite graph, which shows both the most salient entities from each modality and the strengths of their inter-relations.

Since entities of different modalities correlate with each other, they will mutually affect each other's importance in the community. For example, suppose a community contains venue categories: restaurant, cafe, home, and etc, and tip topics: food/drink, hotel, and etc, where the venue category: restaurant has strong correlations with tip topic: food/drink, we should increase the importance level of the venue category: restaurant and tip topic: food/drink to make them more differentiable from the rest of insignificant entities.

We propose an iterative procedure to compute the importance of each entity as follows. Let $\mathbb{U}_{\mathbb{C}}$, $\mathbb{L}_{\mathbb{C}}$, $\mathbb{T}_{\mathbb{C}}$, and $\mathbb{D}_{\mathbb{C}}$ be the sets containing entities from users, venues, tips, and photos of community $\mathbb{C}$, respectively, such that $\mathbb{U}_{\mathbb{C}}, \mathbb{L}_{\mathbb{C}}, \mathbb{T}_{\mathbb{C}}, \mathbb{D}_{\mathbb{C}} \subseteq \mathbb{C}$. We then define the updating function of the importance score of each venue in community $\mathbb{C}$ as:

$$S^{(t+1)}(l, \mathbb{C}) = S^{(t)}(l, \mathbb{C}) \left[\sum_{e \in \mathbb{T}_{\mathbb{C}} \cup \mathbb{D}_{\mathbb{C}}, u \in \mathbb{U}_{\mathbb{C}}} w((u, l, e)) S^{(t)}(e, \mathbb{C}) \right], \qquad (3.10)$$

where $S^{(t)}(l, \mathbb{C})$ is the importance score of venue l in community $\mathbb{C}$ at the tth iteration and $w((u, l, e))$ is the weight of the hyperedge (u, l, e). Similarly, the updating function of the importance score of each tip and photo in community $\mathbb{C}$ is:

$$S^{(t+1)}(e, \mathbb{C}) = S^{(t)}(e, \mathbb{C}) \left[\sum_{l \in \mathbb{L}_{\mathbb{C}}, u \in \mathbb{U}_{\mathbb{C}}} w((u, l, e)) S^{(t)}(l, \mathbb{C}) \right], \qquad (3.11)$$

where $S^{(t)}(e, \mathbb{C})$ is the importance score of entity e being either a tip or a photo in community $\mathbb{C}$ at the tth iteration. Analogous to the TF-IDF concept in text mining, we define the initialization of the entities as:

$$S^{(0)}(e, \mathbb{C}) = -p(e, \mathbb{C}) \sum_{\mathbb{C}' \neq \mathbb{C}} \log p(e, \mathbb{C}'), \qquad (3.12)$$

where $p(e, \mathbb{C})$ is the probability of entity $e \in \mathbb{C}$. We iteratively update the importance score of each entity according to Eqs. (3.10) and (3.11) until the maximum number of iterations is reached, which is set to 500. We then rank entities of each modality according to their final importance scores.

3.7.3.3 Community Visualization and Understanding

We build a tripartite graph, with vertices from multimodality entities (venue categories, tip topics and photo concepts) and edges connecting entities of different modalities to visualize each selected community. The more salient entities (those with highest importance scores) in each modality are showed with bigger size and the stronger inter-entity correlations are represented with thicker edges. With our proposed approach, entities from different modalities are guaranteed to be available to collectively present the profile of each group.

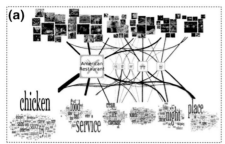
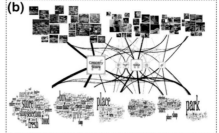

Fig. 3.6 **a** Community of people who enjoy eating with 1,567 members. **b** Community of people who enjoy shopping with 1,108 members. Visualization of two selected representative communities in the global scale (Please view in high 200 % resolution in Acrobat Reader)

We visualize two selected communities (food lovers and shoppers) in Fig. 3.6.[12] We first observe that food lovers visit American restaurants most frequently, which reveals that the majority of the active Foursquare users are located in U.S. Some of them visit bars or night clubs after their meals at the restaurants. The most prominent tip topics posted by food lovers such as "food," "services," "fried chicken," and "night life" correlate well with the venue categories in the community. In addition, we observe that photos related to restaurant, dining, and night club are prominent photo concepts in the community as is demonstrated Fig. 3.6a. Next, we observe that the three most popular venue categories that shoppers visit are the grocery stores, malls, and department stores besides home. Grocery stores mainly retail food, which correlate well with the most prominent discussion topics and photo concepts in the community as demonstrated in Fig. 3.6b. We put the top ten communities detected at the global scale in the online appendix.

While differentiable collective behaviors are exhibited in different communities, we observe that people perform consistent proportion of check-ins at homes and offices across different communities. This pattern reveals the common everyday human behaviors: (home → office → entertainment → home), where the number of check-ins at homes is consistently higher than that at offices.

In additional to the global scale analysis, we next focus on visualizing communities detected at the city scale, where we select communities of similar types from Singapore and New York City and observe some interesting culture differences. Figure 3.7a, b visualize profiles of food lovers in Singapore and New York City, respectively. As expected, food lovers in Singapore often visit Asian Restaurants, Food Courts, and Chinese Restaurants while those in New York City mostly visit American Restaurants. Besides, people in Singapore often discuss topics, such as chicken rice[13] and noodles while those in New York City mostly talk about salad, burger, and fried chicken. We further analyze what are the second popular venues

---

[12] Please refer to the online appendix for the complete list of top ten communities at the global scale.

[13] Chicken rice is one of the famous local delights in Singapore.

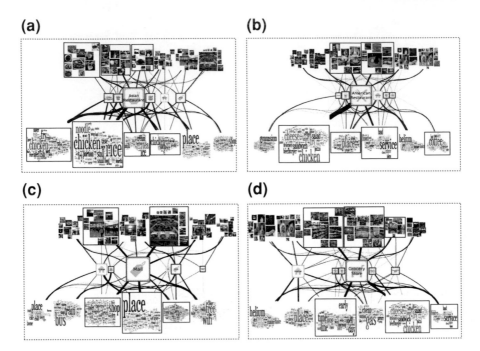

Fig. 3.7 **a** Food lovers in Singapore (679 members). **b** Food lovers in New York City (784 members). **c** Shoppers in Singapore (599 members). **d** Shoppers in New York City (643 members). Comparisons between communities of food lovers/shoppers in Singapore and New York City (*Red rectangles* highlight the most prominent entities while *purple* ones highlight the second prominent groups of categories besides Home) (Please view in high 200 % resolution in Acrobat Reader)

(besides homes) these food lovers visit besides restaurants, where we find that food lovers in Singapore visit malls before or after their meals while those in New York City go to either gyms or offices. These behaviors are as expected, since food courts and many restaurants are usually located in shopping malls in Singapore. Other interesting observations are found by comparing profiles of shoppers in the two cities. As shown in Fig. 3.7c, d, the prominent shopping venues in Singapore are malls and shops, while shoppers in New York City often visit grocery and department stores. More interestingly, most shoppers in Singapore take public transport whereas the counterpart in New York City mostly drive to shop. The tip topics in the two communities also reveal this phenomenon. Besides, some shoppers in Singapore go to some food chains, such as KFC, sandwich shops to take a rest or surf Internet while those in New York City go to coffee shops or restaurants before or after their shopping. The profiles of the top five communities in these two cities are presented in the online appendix.

3.8 Related Work

3.8.1 Human Mobility Analysis in LBSNs

Before the emergence of LBSNs, researchers made use of GPS trajectories to find similar users [40] or infer social ties [41]. However, raw GPS coordinates do not reveal the semantic meanings of locations, thus additional efforts are required to first identify the semantics of each locations before further analysis. Recently, LBSNs are emerging and becoming more and more popular thanks to the recent availability of open mobile platforms, which makes them much more accessible to mobile users. This information provides the opportunity to gain insights on human mobility at unprecedented temporal and user participation scales [3, 13, 24, 25, 34]. Noulas et al. analyzed the user check-in dynamics and the presence of spatio-temporal patterns in Foursquare [24]. More recent work on recommending friends and places in LBSNs by Scellato et al. suggests that the inclusion of information about location-based activities is able to lead to a better prediction than if only social data is considered [29]. Li and Chen used clustering approaches to identify user behavior patterns on BrightKite where they identified five groups of users [13]. Noulas et al. used spectral clustering to group Foursquare users based on the categories of venues they had visited, aiming at characterizing the type of activities of each community [25]. More recently, Vasconcelos et al. grouped Foursquare users into four groups based on the statistical attributes [34]. These studies offer important insights into how users in LBSNs can be grouped based on their interactions, however, the profiles of mined communities are all manually inspected and labeled and the mined communities are only at a very "coarse" level, such as "active users," "normal users," etc. And they did not consider using multimedia information in the process of community detection. Finally, Brown et al. analyzed social and local communities in Gowalla and showed that the two community structures do not yield the same user groupings [3]. However, their focus is different from ours, where we mainly focus on understanding the user communities and characterizing them by means of user-generated multimedia information.

3.8.2 Community Detection

Numerous techniques have been developed for both efficient and effective community detection, including random walks, spectral clustering, modularity maximization, statistical mechanics, etc [8]. In addition, it is well understood that users are naturally characterized by *multiple* community memberships [42]. For example, a

user may be interested in both football and iPad, and thus he/she is very likely
to be a member of these two separated communities. Moreover, it is interesting
and important to interpret and understand the *group profile* of each mined community [32]. Given a complex network with heterogenous entities and interactions,
optimal community detection approaches are expected to be able to mine overlapping
communities which have clear semantic interpretation without restricting the input
network types and requiring prior knowledge on number of latent communities. We
next review community detection approaches which handle input network types in
the order from the simplest to the most complicated.

The simplest network type contains only one type of vertices with one dimension of interactions (Fig. 3.8a). Approaches tackling this kind of networks have
obtained satisfactory results with efficiency and efficacy [22, 38]. However, in
reality, people interact with each other in various forms of activities, which lead
to multi-edge networks among the same set of users, or a *multidimensional network*
with each dimension representing one type of interaction (Fig. 3.8b). For example,
users in LBSNs relate to each other by co-visiting the same venues and by uploading photos with similar visual scenes. Motivated by the modularity maximization
approach [22], Tang et al. proposed principle modularity maximization to identify
the hidden structures shared across dimensions in multidimensional networks [31].
However, Tang's approach is only able to reveal nonoverlapping communities of
users. To discover overlapping communities, Wang et al. proposed a co-clustering
framework, which takes advantage of connection information between users and tags
in social media [37]. While previous work focused on *one-modality* network, a number of emerging applications such as web mining, collaborative filtering, and online
photo sharing involve multiple types of entities and multiple heterogeneous interactions between different types of entities. For example, in social tagging systems,
users collaboratively manage tags to annotate resources and the tagging relationship
involves three different entities: users, resources, and tags. These kinds of *multimodality* network need a very different treatment for community detection tasks
[16, 20] (Fig. 3.8c). Liu and Murata proposed a structural information compression
approach to detect communities in a general k-partite k-uniform hypernetwork [16].
In this chapter, we go one step further by considering community detection in a
k-partite nonuniform hypernetwork, where each hyperedge may involve different
number of vertices from the same/different modalities (Fig. 3.8d). Figure 3.8 gives
a summary of the different network types that the state of the arts have tackled in
the community detection domain and the new network type we aim to tackle in our
work.

3.9 Summary

In this chapter, we investigated the problem of community understanding in LBSNs.
We proposed a novel and unified framework which models heterogenous entities
and interactions by constructing a heterogenous, nonuniform hypergraph. We then

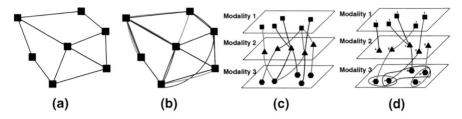

Fig. 3.8 A summary of the different network types. **a** A one-modality single-edge graph. **b** A one-modality multi-edge graph. **c** A tripartite and 3-uniform hypergraph. **d** A heterogeneous and nonuniform hypergraph. In (**d**), there exists the following hyperedges: (1,A), (2,C,III), (3,C), (4,B,I), (5,D,II), (E,IV), (I,II), (II,III,IV), and (IV,V) (Best view in color)

formulated it as a problem to detect dense subgraph over hypergraph, where constraints were added to ensure the interpretability of the detected communities. We then proposed an efficient procedure to solve the optimization problem. Extensive experiments have been performed both qualitatively and quantitatively to verify our proposed approach. Meaningful and interpretable communities were detected in an optimal way while interesting culture differences were revealed by analyzing the communities in Singapore and New York City.

There are a few interesting aspects worth further exploration. First, the time-dependent users' behaviors allow interest communities to be detected and understood in a timely manner. For example, it is interesting to mine and profile different interest groups, which are active during different time periods. Second, users often participate in various social networks. The aggregation of user behaviors across multiple sources is expected to lead to more accurate and timely communities and enrich community understanding.

Appendix

3.10 Top Ten Communities at the Global Scale

Rank	Visualization	Profile
1		• Coffee/Tealovers • GroupSize:1,651
2		• Foodlovers • GroupSize:1,567
3		• Shoppers • GroupSize:1,108
4		• Drink/Snacklovers • GroupSize:1,101
5		• SportsEnthusiasts • GroupSize:1,095

Rank	Visualization	Profile
6		• Home Lovers • Group Size: 1,081
7		• Travellers • Group Size: 997
8		• White-collar Workers • Group Size: 957
9		• Students • Group Size: 933
10		• Nature Lovers • Group Size: 915

3.11 Top Five Communities in Singapore

Rank	Visualization	Profile
1		• Coffee/Tea lovers • Group Size: 795 • Note: In Singapore, some food courts in the neighborhoods are also named coffee shops
2		• Food lovers • Group Size: 679
3		• Shoppers • Group Size: 599
4		• Travellers • Group Size: 584
5		• Fast Food Lovers • Group Size: 576

3.12 Top Five Communities in New York City

Rank	Visualization	Profile
1		• Food lovers • Group Size: 784
2		• Clubbing lovers • Group Size: 698
3		• Shoppers • Group Size: 643
4		• Travellers • Group Size: 639
5		• Tourists • Group Size: 589

References

1. Amitay, E., Carmel, D., Har'El, N., Ofek-Koifman, S., Soffer, A., Yogev, S., Golbandi, N.: Social search and discovery using a unified approach. In: Proceedings of the Conference on Hypertext and Hypermedia, pp. 199–208. ACM (2009)
2. Blei, D.M., Ng, A.Y., Jordan, M.I.: Latent dirichlet allocation. J. Mach. Learn. Res. **3**, 993–1022 (2003)
3. Brown, C., Nicosia, V., Scellato, S., Noulas, A., Mascolo, C.: Social and place-focused communities in location-based online social networks. arXiv, preprint arXiv:1303.6460 (2013)
4. Cao, L., Qi, G.J., Tsai, S.F., Tsai, M.H., Pozo, A.D., Huang, T.S., Zhang, X., Lim, S.H.: Multimedia information networks in social media. Social Network Data Analytics, pp. 413–445. Springer, New York (2011)
5. Dhillon, I.S., Guan, Y., Kulis, B.: Weighted graph cuts without eigenvectors a multilevel approach. IEEE Trans. Pattern Anal. Mach. Intell. **29**(11), 1944–1957 (2007)
6. El-Arini, K., Paquet, U., Herbrich, R., van Gael, J., Agüera y Arcas, B.: Transparent user models for personalization. In: Proceedings of the 18th ACM SIGKDD International Conference on Knowledge Discovery and Data Mining, KDD '12, pp. 678–686. ACM, New York, USA, (2012)
7. Fang, Q., Sang, J., Xu, C., Lu, K.: Paint the city colorfully: location visualization from multiple themes. Advances in Multimedia Modeling. Lecture Notes in Computer Science, vol. 7732, pp. 92–105. Springer, Berlin Heidelberg (2013)
8. Fortunato, S.: Community detection in graphs. Phys. Rep. **486**(3–5), 75–174 (2010)
9. Girvan, M., Newman, M.E.J.: Community structure in social and biological networks. Proc. Nat. Acad. Sci. **99**(12), 7821 (2002)
10. Guimera, R., Sales-Pardo, M., Amaral, L.A.N.: Modularity from fluctuations in random graphs and complex networks. Phys. Rev. E **70**(2), 025101 (2004)
11. Ken Chatfield, A.V., Lempitsky, V., Zisserman, A.: The devil is in the details: an evaluation of recent feature encoding methods. In: Hoey, J., McKenna, S., Trucco, E. (eds.) Proceedings of the British Machine Vision Conference, pp. 76.1-76.12. BMVA Press (2011)
12. Leskovec, J., Lang, K.J., Mahoney, M.: Empirical comparison of algorithms for network community detection. In: Proceedings of the 19th International Conference on World Wide Web, WWW '10, pp. 631–640. ACM, New York, USA (2010)
13. Li, N., Chen, G.: Analysis of a location-based social network. In: International Conference on Computational Science and Engineering, 2009. CSE '09, vol. 4, pp. 263–270 (2009)
14. Lin, Y.-R., Sundaram, H., de Choudhury, M., Kelliher, A.: Discovering multirelational structure in social media streams. ACM Trans. Multimedia Comput. Commun. Appl. **8**(1), 4:1–4:28 (2012)
15. Liu, H., Latecki, L.J., Yan, S.: Robust clustering as ensembles of affinity relations. In: Proceedings of Advances in Neural Information Processing Systems(2010)
16. Liu, X., Murata, T.: Detecting communities in k-partite k-uniform (hyper) networks. J. Comput. Sci. Technol. **26**(5), 778–791 (2011)
17. Lu, C., Hu, X., Park, J.R.: Exploiting the social tagging network for web clustering. IEEE Trans. Syst. Man Cybern. Part A Syst. Hum. **41**(5), 840–852 (2011)
18. McCallum, A.K.: Mallet: a machine learning for language toolkit. http://mallet.cs.umass.edu (2002)
19. Mei, T., Li, L., Hua, X.-S., Li, S.: Imagesense: towards contextual image advertising. ACM Trans. Multimedia Comput. Commun. Appl. **8**(1), 6:1–6:18 (2012)
20. Murata, T., Ikeya, T.: A new modularity for detecting one-to-many correspondence of communities in bipartite networks. Adv. Complex Syst. **13**(1), 19–31 (2010)
21. Neubauer, N., Obermayer, K.: Towards community detection in k-partite k-uniform hypergraphs. In: Proceedings of the NIPS Workshop on Analyzing Networks and Learning with Graphs (2009)
22. Newman, M.E.J.: Finding community structure in networks using the eigenvectors of matrices. Phys. Rev. E **74**(3), 036104 (2006)

23. Nie, W., Wang, X., Zhao, Y.-L., Gao, Y., Su, Y., Chua, T.-S.: Venue semantics: multimedia topic modeling of social media contents. In: Huet, B., Ngo, C.-W., Tang, J., Zhou, Z.-H., Hauptmann, A.G., Yan, S. (eds.) Advances in Multimedia Information Processing C PCM 2013, Lecture Notes in Computer Science, vol. 8294, pp. 574–585. Springer, New York (2013)
24. Noulas, A., Scellato, S., Mascolo, C., Pontil, M.: An empirical study of geographic user activity patterns in foursquare. In: Fifth International AAAI Conference on Weblogs and Social Media (2011)
25. Noulas, A., Scellato, S., Mascolo, C., Pontil, M.: Exploiting semantic annotations for clustering geographic areas and users in location-based social networks. In: Proceedings of the Workshop Social Mobile Web (2011)
26. Papadopoulos, S., Kompatsiaris, Y., Vakali, A., Spyridonos, P.: Community detection in social media. Data Min. Knowl. Disc. 24(3), 515–554 (2012)
27. Perronnin, F., Sánchez, J., Mensink, T.: Improving the fisher kernel for large-scale image classification. In: Proceedings of the 11th European Conference on Computer Vision: Part IV, ECCV'10, pp. 143–156. Springer, Berlin (2010)
28. Porter, M.A., Onnela, J.-P., Mucha, P.J.: Communities in networks. Not. AMS 56(9), 1082–1097 (2009)
29. Scellato, S., Noulas, A., Mascolo, C.: Exploiting place features in link prediction on location-based social networks. In: Proceedings of the 17th ACM SIGKDD International Conference on Knowledge Discovery and Data Mining, KDD '11, pp. 1046–1054. ACM, New York, USA (2011)
30. Tang, L., Liu, H.: Scalable learning of collective behavior based on sparse social dimensions. In: Proceedings of the 18th ACM Conference on Information and Knowledge Management, CIKM '09, pp. 1107–1116. ACM, New York, USA (2009)
31. Tang, L., Wang, X., Liu, H.: Uncoverning groups via heterogeneous interaction analysis. In: Proceedings of the 2009 Ninth IEEE International Conference on Data Mining, ICDM '09, pp. 503–512. IEEE Computer Society, Washington, USA (2009)
32. Tang, L., Wang, X., Liu, H.: Understanding emerging social structuresa group profiling approach. School of Computing, Informatics, and Decision Systems Engineering, Arizona State University, Tech. Rep. TR-10-002 (2010)
33. Van, M.E.L., Zisserman, A.: The pascal visual object classes (voc) challenge. Int. J. Comput. Vis. 88(3), 3–338 (2010)
34. Vasconcelos, M.A., Ricci, S., Almeida, J., Benevenuto, F., Almeida, V.: Tips, dones and todos: uncovering user profiles in foursquare. In: Proceedings of the Fifth ACM International Conference on Web Search and Data Mining, WSDM '12, pp. 653–662. ACM, New York, USA (2012)
35. Vedaldi, A., Fulkerson, B.: VLFeat: An open and portable library of computer vision algorithms. http://www.vlfeat.org/ (2008)
36. Wang, M., Ni, B., Hua, X.-S., Chua, T.-S.: Assistive tagging: a survey of multimedia tagging with human-computer joint exploration. ACM Comput. Surv. 44(4), 25:1–25:24 (2012)
37. Wang, X., Tang, L., Gao, H., Liu, H.: Discovering overlapping groups in social media. In: Proceedings of the 2010 IEEE International Conference on Data Mining, ICDM '10, pp. 569–578. IEEE Computer Society, Washington, USA (2010)
38. Wolfe, A.W.: Social network analysis: methods and applications. Am. Ethnologist 24(1), 219–220 (1997)
39. Xiao, J., Hays, J., Ehinger, K.A., Oliva, A., Torralba, A.: Sun database: large-scale scene recognition from abbey to zoo. In: IEEE Conference on Computer Vision and Pattern Recognition (CVPR), 2010, pp. 3485–3492 (2010)
40. Xiao, X., Zheng, Y., Luo, Q., Xie, X.: Finding similar users using category-based location history. In: Proceedings of the 18th SIGSPATIAL International Conference on Advances in Geographic Information Systems, pp. 442–445. ACM (2010)
41. Xiao, X., Zheng, Y., Luo, Q., Xie, X.: Inferring social ties between users with human location history. J. Ambient Intell. Hum. Comput. 5(1), 3–19 (2014)

42. Xie, J., Kelley, S., Szymanski, B.K.: Overlapping community detection in networks: the state of the art and comparative study. Arxiv, preprint arXiv:1110.5813 (2011)
43. Xu, B., Bu, J., Chen, C., Cai, D.: An exploration of improving collaborative recommender systems via user-item subgroups. In: Proceedings of the International Conference on World Wide Web, pp. 21–30. ACM (2012)
44. Zhao, Y.-L., Zheng, Y.-T., Zhou, X., Chua, T.-S.: Generating representative views of landmarks via scenic theme detection. In: Proceedings of the International Conference on Advances in Multimedia Modeling—Volume Part I, pp. 392–402. Springer-Verlag (2011)
45. Zheng, V.W., Zheng, Y., Xie, X., Yang, Q.: Collaborative location and activity recommendations with gps history data. In: Proceedings of the 19th International Conference on World Wide Web, WWW '10, pp. 1029–1038. ACM, New York, USA (2010)
46. Zhou, T., Ren, J., Medo, M., Zhang, Y.C.: Bipartite network projection and personal recommendation. Phys. Rev. E **76**(4), 046115 (2007)
47. Zhuang, J., Mei, T., Hoi, S.C.H., Xu, Y.-Q., Li, S.: When recommendation meets mobile: contextual and personalized recommendation on the go. In: Proceedings of the 13th International Conference on Ubiquitous Computing, UbiComp '11, pp. 153–162. ACM, New York, USA (2011)

Chapter 4
Social Role Recognition for Human Event Understanding

Vignesh Ramanathan, Bangpeng Yao and Li Fei-Fei

4.1 Introduction

Humans are social animals. Our ability to comprehend human relations stands fundamental to our survival, development, and social life. Inspired by the observation that social relationship is an important cue in understanding events and activities, we set out to examine the problem of assigning social roles to different characters involved in the video of a social event such as a wedding or a birthday party. While this problem is closely related to person detection in images or videos, we emphasize on the relatively novel task of role recognition.

We understand human relationships in terms of social roles assumed by people, and tend to describe events using these roles. For instance, consider the video in Fig. 4.1. A perfect action recognition system would only provide us the descriptions shown in Fig. 4.1a. However, we would prefer a more compete desciption involving the roles played by each person as shown in Fig. 4.1b. Typically, social roles answer semantic queries like, "Who is doing what in an event?" While the tasks of identifying the action and detecting the person (depicted in Fig. 4.1a) are widely studied in computer vision, the problem of role assignment (depicted in Fig. 4.1b) is relatively new and equally interesting. This would lead to a richer description of human events, as a next step in video understanding.

V. Ramanathan (✉)
Department of Electrical Engineering, Stanford University, 119 Quillen Ct
1113, Stanford, CA 94305, USA
e-mail: vigneshr@cs.stanford.edu; vigneshram.iitkgp@gmail.com

B. Yao · L. Fei-Fei
Computer Science Department, Stanford University, Stanford, USA
e-mail: bangpeng.yao@cs.stanford.edu

L. Fei-Fei
e-mail: feifeili@cs.stanford.edu

Y. Fu (ed.), *Human-Centered Social Media Analytics*,
DOI: 10.1007/978-3-319-05491-9_4, © Springer International Publishing Switzerland 2014

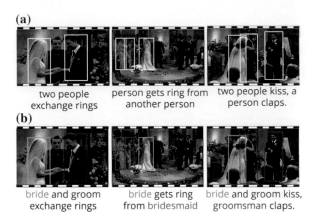

Fig. 4.1 **a** Description of a video only based on actions is shown. The relevant people are enclosed in a *bounding box*. **b** A more complete description involving social roles is shown. The *color* of the *bounding box* corresponds to the role mentioned in the description. When people interact in an event, they assume event-specific social roles. Social roles act as identities for the individuals and can help us describe the event in terms of these roles

Social role discovery derives motivation from the field of "Role Theory" [2] in sociology, which defines social roles as cultural objects "used to accomplish pragmatic interaction goals in a community," i.e., roles are decided on the basis of human interactions with each other and the event environment. This shows that knowing the role of a person can help determine his/her interactions with the environment and vice versa. In computer vision, Lan et al. [15] leveraged the same intuition to build a human activity recognition model. Also, the knowledge of social roles can help determine the interesting segments of social event footages [9] and sports videos.

The definition of social roles is event specific, and can sometimes be abstract such as, people "helping," "visiting," or "residing" in a nursing home [15], making role identification a difficult human task. Ideally, we would like to automatically discover such interaction-based role assignments in any event. Also, annotating roles is time-consuming and needs knowledge of the event. Recognizing these difficulties, we formulate the problem of social role discovery in a weakly supervised framework. Given a set of videos belonging to a social event without training labels for the people in the videos, we group them into different social roles. The event label acts as the weak annotation in our setting, restricting the discovered roles to be event specific.

The problem is amply challenging due to the wide variation in appearance, scale, location, and scene context of a role across different videos as seen in Fig. 4.2. Further, it is difficult to determine roles just by observing people individually. Rather, social role discovery is an attempt to identify people based on their interactions in an event. Modeling such interactions in the absence of role labels during training acts as an additional challenge.

In order to solve this problem of weakly supervised role assignment, we propose a Conditional Random Field (CRF) to capture inter-role interaction cues, and develop

Fig. 4.2 Sample frames from different events in the YouTube social roles dataset are shown with ground truth role annotations used for evaluation. The different roles in each event are marked by the *colors* noted in the last column. The huge variation in appearance, location, scale, and scene context for a role across different videos can be seen

a tractable variational inference procedure to jointly learn role labels as well as model weights in [23]. Further, to evaluate the model performance, we introduce a novel YouTube social roles dataset in Sect. 4.5.1, accompanied by event-specific ground truth role annotations for the people in the videos. It is to be noted that the role labels are only used for model evaluation and not for the training. This enables our method to automatically discover the human relationships in an event, and removes the need for tedious role annotation in videos. Hence, our method clusters people from videos of the same event-class into different social groups. Since we do not use any labels to train the models, the role labels associated with these social groups are not provided by our method. In practice, if the labels are known for a few individiuals in a social group, it can be propogated to others in the group. We also provide role annotations for a subset of videos from two events of the TRECVID MED-11 [1] event kits, and test our model performance on these videos. Experiments on these datasets show that our method achieves encouraging performance in weakly supervised social role assignment.

4.2 Related Work

Socially aware video and image analysis. Recent works on social network construction and interaction understanding is relevant to our work on social role recognition. Yu et al. [30] associate people in a video using face recognition and track matching.

Ding and Yilmaz [6, 7] cluster people in a movie into adversarial groups. Ding and Yilmaz [7] use scene context and visual concept attributes to build social relation network. Weng et al. [28] also build a social role network -based on their co-occurrence of movie characters in different scenes. These works do not group people across different videos, but consider people within one movie. Wang et al. [27] use appearance features to predict the relationship between people by training on images with weak relationship labels, while Song et al. [24] perform occupation classification based on clothing and context in human images. Stone et al. [25] studied the problem of face recognition in social context.

Identifying Social Interactions.   One of the first datasets aimed at studying social interactions was introduced in [3]. This is a dataset of pairs of mice engaged in different social behaviors, categorized into different classes like "attack," "copulation," and "chase." They identified these behaviors by segmenting video sequences based on trajectory features extracted from mouse tracks. Cristani et al. [5] focused on outdoor scenarios, where they identify different formations of humans in a social event through a voting strategy. More recently, Li et al. [18] introduced a database of students in a classroom setting. In a cluttered social setting, they identify the video segments corresponding to interactions like "hug," "shake," and "fight" based on Metric Learning. These works provide a better way of understanding participant behavior in a crowded setting.

Modeling Interactions for Action Recognition.   Another related line of work has been the use of social interaction to aid group action recognition [4, 8, 16]. Lan et al. [16] identify human interactions and [4] use features of people in spatiotemporal vicinity to detect group activities and jointly track multiple people. Qin and Shelton [22] also use social grouping to help multi target tracking. The above works group people primarily based on their spatial proxmity and do not generalize to social events with complex interactions. Unlike these methods, we focus on grouping people into social roles based on richer interaction features in any human event.

Gallagher and Chen [12] use social context in group photos to make better prediction of human attributes and scene semantics. Fu et al. [11] recognize group social activities through attribute learning. Perez et al. [21] develop interaction features based on facial orientation to recognize activities like hand-shaking. Similarly, Marin-Jimenez et al. [20] also model facial attention. While these works make effective use of social cues to aid different tasks, they do not explicitly categorize people based on their roles in a social event.

Social Role Recognition.   Recently, Fathi et al. [9] and Lan et al. [15] used social roles to predict group activities. Fathi et al. [9] found face attention patterns in first-person videos to detect interaction activities like monolog, discussion, and dialog. They clustered faces in training videos based on attention patterns, and represented frame sequences by histogram of cluster occurrences. Lan et al. [15] predicted role labels like "defender" and "attacker" in sports videos to identify group activities. They used training labels to learn role assignments based on spatiotemporal interaction between players. They showed a significant improvement in action and event recognition by modeling social roles, actions, and events in a hierarchical setting. This can again be attributed to the association of specific human tasks with certain roles (like attacker running, goalkeeper catching the ball), as postualted by Role Theory. This was also established by their observation of strong correlations between certain actions and roles. However, in our work we are not provided role annotations, and we wish to discover interaction-based roles automatically by studying different instances of an event. We also use richer interaction features.

4.3 Our Approach

We define social role discovery as a weakly supervised problem, where the training role labels for the people in the videos are not available. We are only provided the event label for each video, and the number of roles to be discovered in an event. In our work, we focus on the grouping of people into different social clusters based on their interactions. We assume that the people in the video are spatiotemporally localized. The spatiotemporal location of a person is referred to as the *track* of the person. We assume that every video is preprocessed to obtain individual human tracks similar to [8, 15]. In our experiments, the tracks were obtained through the active learning tool provided in [26].

Social roles are not only decided by person-specific descriptors, but also by the interaction between people. Hence, any model used to discover social roles should be capable of incorporating this information. However, interaction in an event is usually restricted to a small set of roles. Every social event has a *reference role* consistently appearing across all instances of the event. This reference role provides the most useful information regarding the event, as well as other roles in the event. To understand this, consider a *birthday*, where the important interactions mostly involve the "birthday person." Discovering the reference role in a video is an interesting problem in itself and can answer semantic queries like "Who's birthday is it?" or "Who is teaching the class?" We wish to identify the reference role in different videos of an event and simultaneously group other people across all these videos into meaningful social roles. With this assumption of a reference role, it is sufficient to model the interaction of any person only with the reference role. This is a realistic simplification, enabling us to perform tractable inference as shown in Sect. 4.4. One instance of the reference role is assumed to be present in every video belonging to the event class. We refer to the other roles as secondary roles.

4.3.1 Model Formulation

We present a CRF model which accounts for the reference role interaction with other roles in a video. An overview of our approach is shown in Fig. 4.3, along with the factor graph of our model. As illustrated, to capture person-specific social cues, we extract unary features (Ψ_u) from each human track, describing spatiotemporal

Fig. 4.3 a The features extracted by our model are illustrated on a sample birthday video frame (shown in the *left*). Unary features are represented in *blue*, while the pairwise features are shown in *red*. **b** The factor graph of our CRF model is shown. The observed variables are *shaded*. m is the index of the reference role in the video v. The model variables are as defined in Sect. 4.3.1. *Note For color interpretation see online version*. **c** The CRF model is illustrated on the birthday video

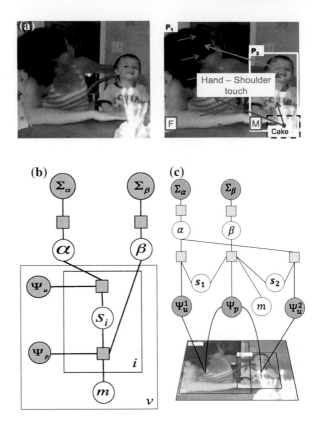

activity, human appearance, and human–object interaction. Similarly, to represent interaction-based social cues, pairwise features (Ψ_p) describing proxemic touch codes and spatial proximity are extracted. Our CRF model uses these features to perform weakly supervised social role recognition.

Let $\mathbb{P}_v$ be the set of people in a video v and s_i^v be the social role assigned to a person $p_i^v \in \mathbb{P}_v$. We want to assign social roles, and jointly learn model weights by maximizing the log likelihood of the CRF shown in Eq. (4.1)

$$\underset{s_E,\alpha,\beta}{\operatorname{argmax}} \sum_v \left\{ \sum_{p_i^v} \alpha \cdot \Psi_u(p_i^v, s_i^v) + \sum_{p_j^v \neq p_m^v} \beta \cdot \Psi_p(p_m^v, p_j^v, s_j^v) - Z_v \right\} \quad (4.1)$$
$$- \frac{\alpha^T \Sigma_\alpha^{-1} \alpha + \beta^T \Sigma_\beta^{-1} \beta}{2},$$

where m_E denotes the reference role in the event E, and p_m^v the person holding the reference role in v. The model potentials are defined as

$$\alpha \cdot \Psi_u(p_i^v, s_i^v) = \sum_s \alpha_s \cdot \mathbf{1}(s = s_i^v)\Psi_u(p_i^v), \qquad (4.2)$$

$$\beta \cdot \Psi_p(p_m^v, p_j^v, s_j^v) = \sum_{s \neq m_E} \beta_s \cdot \mathbf{1}(s = s_j^v)\Psi_p(p_m^v, p_j^v)$$

In Eq. (4.1), s_E is the complete social role assignment to all people in the event, and Z_v is the log-partition function for the video v. Σ_α, and Σ_β are the covariances of the Gaussian priors on α and β respectively. Note that the model only considers interaction of different roles with the reference role, in accordance with our assumption, and every video is assumed to contain one person playing this reference role. α and β are the unary and pairwise weights to be learned respectively. A factor graph of the model is shown in Fig. 4.3.

4.3.2 Unary Features

The unary feature Ψ_u captures role-specific social cues extracted from human tracks, and their interaction with the event environment. Ψ_u can be expanded into four components as shown below.

Histogram of Gradient Feature Ψ_u^{HoG}: Bag of densely computed HoG3D [13] words of dimension 1,429 along the human track is used as low-level features to capture the individual actions.

Spatiotemporal Feature Ψ_u^{ST}: A person's movement in an event is another useful cue regarding his/ her role. For example, the "bride" often walks down the aisle in a church *wedding*. The human motion between two frames is binned along eight directions to form a trajectory feature similar to [14]. These features are normalized across different people in a video to partly account for camera motion.

Object Interaction Feature Ψ_u^{OI}: The interaction of a person with the event environment plays a key role in determining his/ her role. "birthday person" cutting a "cake" and "function host" talking at the "lectern" are representative examples. In the current work, we extract interaction features corresponding to only these two objects in the respective events. Felzenszwalb et al. [10] is used to obtain specific object detection scores in a video. These scores are spatially pooled similar to [17] in the periphery of the person's bounding box and averaged across multiple frames to form an object interaction feature of dimension 48 for every event object.

Social Feature $\Psi_u^{Soc.}$: These features capture two important social aspects of a person, representing gender and clothing. Such cues are important in events like *wedding*. This would also capture the gender bias in certain roles like "brides." We first use [32] to detect faces and obtain scores [1] for gender classification. The scores are averaged across frames to form the gender feature. We pool the RGB values in the upper half of the human bounding box to form a 32 bin histogram feature. This feature is used to represent the clothing of the person.

---

[1] We use software from http://cmp.felk.cvut.cz/~fisarond/demo/

4.3.3 Pairwise Interaction Features

Human interaction forms an important basis for social role definitions. For instance, the "parent" in a *birthday* is distinguished from "guests" by their interaction with the "birthday person." Similarly "bride-groom," "instructor–student" interactions separate the respective roles from others. These interactions are recorded by the pairwise feature Ψ_p composed of two components as shown below.

Proxemic Interaction Feature $\Psi_p^{\text{Prox.}}$: The proxemic interaction of two people provides interesting insights regarding the relation between roles in an event such as the touch-code between a "parent" and the "birthday child." The use of proxemics for describing human–human relations was introduced in [29], where the authors classify proxemics between two people into 6 classes with 20 models. Proxemics are also referred as touch-codes, indicating the way people touch each other. For every pair of humans in a video, we use all 20 models from [29] to find proxemic scores in different frames. The scores are normalized across all human pairs in a given video and split into 16 bins for every model, to form our final proxemic descriptor. The scores are set to a minimum value, if a pair of people are never sufficiently close to each other.

Spatiotemporal Interaction Feature Ψ_p^{ST}: The spatial separation of people across time is a simple but powerful measure of human interaction in a video. For instance, the "bride" and "groom" are always near each other in a wedding, while the "grooms-men" are farther away from the "bride." The spatial distance between a pair, normalized by bounding box dimensions at different time instants are used.

4.4 Inference

The difficulty of solving Eq. (4.1) arises due to the correlation between different social roles and the coupling introduced by Z_v. Zhu and Xing [31] proposed a mean field approximation to solve Conditional Topic Random Fields, with simple chain connected CRFs and CRFs without interaction potentials. Along similar lines, we develop a variational inference method to find an approximate solution for our graphical model. We show that the simplifying assumption of interactions being restricted to the reference role, helps us perform tractable inference as a part of the optimization procedure. We also introduce a variational approximation to the social role probability distribution in a video, with similar dependencies as the original model.

We formulate the variational approximation q of the model distribution as shown in Eq. (4.7), where s_v denotes the role assignment to all people in the video v.

$$q(\alpha, \beta, s_E | \lambda_\alpha, \lambda_\beta, \sigma_\alpha^2, \sigma_\beta^2, \phi, \psi) \tag{4.3}$$
$$= \prod_j q(\alpha^j | \lambda_{\alpha^j}, \sigma_{\alpha^j}^2) \prod_k q(\beta^k | \lambda_{\beta^k}, \sigma_{\beta^k}^2) \prod_v q(s^v | \phi^v, \psi^v)$$

The distributions over α and β are approximated by univariate normal distribution with means given by λ_α, λ_β and variances σ_α^2, σ_β^2. ϕ^v is a factor giving the probability of a person being assigned the reference role in the video. ψ^v is a set of $|\mathbb{P}_v|$ factors, where $\psi_{(i)}^v$ is the secondary role probability matrix for other people in the video, when p_i^v is assigned the reference role. ϕ, ψ are formally defined in Eq. (4.4). This variational approximation of the social role probability, retains the dependencies in our original structure. It represents one predominant reference role, with secondary role assignments dependent on this reference role.

$$\phi^v(p_i^v) = p(s_i^v = m_E) \tag{4.4}$$
$$\psi_{(i)}^v(p_j^v, s) = p(s_j^v = s | s_i^v = m_E), \quad j \neq i, \; s \neq m_E$$

Inference is then carried out through coordinate ascent. In each iteration, the updates for ϕ, ψ require inference in the CRF model, with the model weights fixed. When the model weights are fixed, our graph reduces to a tree for each individual video, allowing us to perform exact clique-tree inference. The optimization procedure and update equations for ψ, ϕ, λ, σ^2 are shown in Sect. 4.4.1. We use the L-BFGS algorithm from [19] to perform gradient ascent in each iteration to update λ, σ^2.

We initialize both ϕ^v, $\psi_{(i)}^v$ to be uniform for all people in the event. λ_{α_s} are initialized to be the maximally separated points in the unary feature space for an event E. λ_{β_s} are similarly initialized from the pairwise interaction feature space. $\sigma_{\alpha^j}^2$ are initialized to 0.01 or 0.1 based on the variance of the event unary features. Similarly, $\sigma_{\beta^k}^2$ are initialized to 10 or 0.1 for all events.

In every video v, the person p_m^v with the highest value of $\phi^v(p_m^v)$ is assigned the reference role. The corresponding variational probability $\psi_{(m)}^v$ is then used to assign secondary roles to other people in the video. While assigning secondary roles, we enforce a lower l and upper u bound on the number of people assigned a secondary role s in the event. In practice, the bounds are set to a 10 % range of the smallest and largest ground truth cluster sizes in the event. This acts as a loss prior to the number of people in each role cluster. Let $\mathbb{P}_{-m_E}$ be the set of people not assigned the reference role in event E. Let ψ be the secondary role probability matrix, where each row corresponds to a person $p_k \in \mathbb{P}_{-m_E}$ and each column represents a secondary role s. ψ can be obtained by stacking $\psi_{(m)}^v$ from all videos. Y is the secondary role assignment matrix with the same dimensions as ψ, where an entry Y_{ks} is set to 1 if the person p_k is assigned the role s. Secondary role assignment is then carried out by solving the linear integer program in Eq. (4.5) to maximize the probability of role assignment under given constraints.

$$\max_{Y} \quad Trace\left(Y^T \psi\right),$$
$$\text{subject to } Y\mathbf{1} = \mathbf{1}, \qquad\qquad\qquad (4.5)$$
$$l \leq Y^T \mathbf{1} \leq u,$$
$$Y_{ks} \in \{0, 1\} \,\forall\, p_k \in \mathbb{P}_{-m_E}$$

We enforce the same constraints in our baseline models as well.

4.4.1 Update Equations for Variational Inference

A lower bound on the log likelihood of the CRF can be derived using Jensen's inequality as shown in Eq. (4.6).

$$\log p\left(s_E, \alpha, \beta | \Sigma_\alpha, \Sigma_\beta\right) \geq \mathbb{L}(q, \Psi_u, \Psi_p) \qquad\qquad (4.6)$$
$$= E_q\left[\log p(\alpha, \beta | \Sigma_\alpha, \Sigma_\beta)\right]$$
$$+ \sum_v E_q\left[\log p(s^v | \alpha, \beta)\right] + H,$$

where s^v is the complete role assignment to all people in the video v, and H is the entropy of the variational distribution q shown in Eq. (4.7).

$$q(\alpha, \beta, s_E | \lambda_\alpha, \lambda_\beta, \sigma_\alpha^2, \sigma_\beta^2, \phi, \psi) \qquad\qquad (4.7)$$
$$= \prod_j q(\alpha^j | \lambda_{\alpha^j}, \sigma_{\alpha^j}^2) \prod_k q(\beta^k | \lambda_{\beta k}, \sigma_{\beta k}^2) \prod_v q(s^v | \phi^v, \psi^v)$$

Now, $E_q\left[\log p(s^v | \alpha, \beta)\right]$ in Eq. (4.6) can be expanded as

$$E_q\left[\log p(s^v | \alpha, \beta)\right] = \lambda_\alpha \cdot E_q\left[\Psi_u(s^v)\right] \qquad\qquad (4.8)$$
$$+ \lambda_\beta \cdot E_q\left[\Psi_p(s^v)\right] - E_q\left[Z_v\right],$$

where, $Z_v = \log\left\{\sum_{s^v} \exp\left(\alpha \cdot \Psi_u(s^v) + \beta \cdot \Psi_p(s^v)\right)\right\}$ is the log partition function. Using the fact, $\log x \leq a^{-1}x - 1 + \log a$, we can establish a lower bound on $E_q\left[\log p(s^v | \alpha, \beta)\right]$ as shown below

$$E_q\left[\log p(s^v | \alpha, \beta)\right] \geq \lambda_\alpha \cdot E_q\left[\Psi_u(s^v)\right] \qquad\qquad (4.9)$$
$$+ \lambda_\beta \cdot E_q\left[\Psi_p(s^v)\right] - \frac{h_v(q)}{\zeta_v} - log(\zeta_v),$$

where ζ_v is a variational parameter and $h_v(q)$ is defined as

$$h_v(q) = \sum_{s^v} E_q \left[\exp\left\{ \alpha \cdot \Psi_u(s^v) + \beta \cdot \Psi_p(s^v) \right\} \right] \tag{4.10}$$

$$= \sum_{s^v} \exp\left\{ \sum_{p_i^v} \lambda_\alpha \cdot \Psi_u(p_i^v, s_i^v) + \frac{\sigma_\alpha^2}{2} \cdot \Psi_u^2(p_i^v, s_i^v) \right.$$

$$\left. + \sum_{p_j^v \neq p_m^v} \lambda_\beta \cdot \Psi_p(p_m^v, p_j^v, s_j^v) + \frac{\sigma_\beta^2}{2} \cdot \Psi_p^2(p_m^v, p_j^v, s_j^v) \right\}$$

Given Σ_α, Σ_β, we update the parameters through a coordinate ascent method to maximize the lower bound in Eq. (4.6). ζ_v is updated to $h_v(q)$ at each iteration. The closed-form update equations for $\phi^v(p_i^v)$, $\psi_{(i)}^v(p_j^v, s)$ at each iteration are shown in Eq. (4.11).

$$\phi^v(p_i^v) \propto \exp\left\{ \lambda_{\alpha_{m_E}} \cdot \Psi_u(p_i^v) \right.$$

$$\left. + \sum_{j \neq i} \sum_{s \neq m_E} \psi_{(i)}(p_j^v, s) \left[\lambda_{\beta_s} \cdot \Psi_p(p_j^v, p_i^v) \right] \right\} \tag{4.11}$$

$$\psi_{(i)}^v(p_j^v, s) \propto \exp\left\{ \phi^v(p_i^v) \lambda_{\alpha_s} \cdot \Psi_u(p_j^v) \right.$$

$$\left. + \phi^v(p_i^v) \left[\lambda_{\beta_s} \cdot \Psi_p(p_i^v, p_j^v) \right] \right\}$$

At each iteration, the values of λ_α, λ_β, and σ_α^2, σ_β^2 are updated using L-BFGS. The gradients of $\mathbb{L}$ with respect to λ_{α_s} and λ_{β_s} are given below

$$\nabla_{\lambda_{\alpha_s}} \mathbb{L} = \Sigma_{\alpha_s}^{-1} \lambda_{\alpha_s} + \sum_v \left\{ \sum_i E_q[\Psi_u(p_i^v, s)] \right. \tag{4.12}$$

$$\left. - \zeta_v^{-1} \nabla_{\lambda_{\alpha_s}} h_v(q) \right\}$$

$$\nabla_{\lambda_{\beta_s}} \mathbb{L} = \Sigma_{\beta_s}^{-1} \lambda_{\beta_s} + \sum_v \left\{ \sum_{\substack{i,j \\ j \neq i}} E_q[\Psi_p(p_i^v, p_j^v, s)] \right.$$

$$\left. - \zeta_v^{-1} \nabla_{\lambda_{\alpha_s}} h_v(q) \right\},$$

where Σ_{α_s}, Σ_{β_s} are the components of Σ_α, Σ_β corresponding to α_s, β_s respectively. As before, $\Psi_p(p_i^v, p_j^v, s)$ is the pairwise feature when p_i^v is the reference role and

p_j^v is assigned the role s. The gradients of $\mathbb{L}$ with respect to $\sigma_{\alpha^k}^2$ and $\sigma_{\beta^k}^2$ are given below

$$\nabla_{\sigma_{\alpha^k}^2} \mathbb{L} = -\frac{1}{2}\Sigma_{\alpha^k}^{-1} - \sum_v \zeta_v^{-1} \nabla_{\sigma_{\alpha^k}^2} h_v(q) + \frac{1}{2\sigma_{\alpha^k}^2} \qquad (4.13)$$

$$\nabla_{\sigma_{\beta^k}^2} \mathbb{L} = -\frac{1}{2}\Sigma_{\beta^k}^{-1} - \sum_v \zeta_v^{-1} \nabla_{\sigma_{\beta^k}^2} h_v(q) + \frac{1}{2\sigma_{\beta^k}^2},$$

where Σ_{α^k} and Σ_{β^k} are the kth diagonal elements in Σ_α and Σ_β respectively.

It is to be noted that the assumption of significant interaction only with the reference role, helps us exactly compute $h_v(q)$, $\nabla_\lambda h_v(q)$, $\nabla_{\sigma^2} h_v(q)$ through a clique-tree messgae passing algorithm. The exact computation of $h_v(q)$ is intractable in a fully connected graph with interaction among all social roles.

Implementation details. ζ_v is initialized to $1E6$ in our experiments. Also, the hyperparameters Σ_α and Σ_β are diagonal matrices whose nonzero entries are all set to $0.01, 0.1$, or 10 based on the variance of the unary and pairwise features. This is indicative of the amount of variance in the respective features.

4.5 Experiment and Results

4.5.1 Datasets

YouTube Social Roles. Most publicly available video datasets are not suitable for evaluating the social role assignment task, since they do not cover a good range of people performing different roles in specific social events. In an attempt to evaluate our method, we collected a set of YouTube videos under 4 social events. The details of the dataset are shown in Table 4.1. To facilitate easy evaluation, we annotate every person in our dataset with the relevant social roles. Some videos have stray individuals not annotated with any specific social role and are called as "others." Again it is to be noted that role labels are used only for evaluation.

Within each social event, there is wide variation in event settings as seen from the sample video frames in Fig. 4.2. *Wedding* and *Birthday* videos were chosen to cover both indoor and outdoor celebrations. *Award ceremony* includes graduation functions, presidential award functions, as well as corporate events. Similarly, *physical training* refers to martial arts, aerobics, and other forms of fitness classes. This diversity in scenarios, with the same underlying interactions between different roles is an interesting characteristic of the dataset, and makes the task amply challenging. *TRECVID Social Roles.* Among publicly available datasets, the TRECVID-MED11 event kits [1] have two social event classes *birthday* and *wedding*. However, most of the videos in these kits either have very few characters or crowd activities where

Table 4.1   Details of the YouTube social roles dataset

Event name	Social roles (no. of people per role)	No. of videos	Avg. duration (s)
Birthday party	Birthday child (40), parents (44), friends (71), guests (28)	40	80.84
Catholic wedding	Bride (40), groom (40), priest (38), grooms men (45), brides maids (43), others (8)	40	88.74
Award function	Presenter (40), receipient (309), host (25), disributor (17), others (13)	40	111.13
Physical training	Instructor (36), students (127)	36	50.49

Table 4.2   Details of the TRECVID social roles dataset

Event name	Social roles (no. of people per role)	No. of videos	Avg. duration (s)
Birthday party	Birthday person (34), parent/spouse (40), friends (59), guests (31)	34	44.65
Catholic wedding	Bride (34), groom (34), priest (29), grooms men (29), brides maids (29)	34	72.00

people cannot be distinguished from each other. Hence, we chose a smaller subset, covering reasonable number of people in different roles. Some videos were cropped to include only the parts showing relevant social events. Details of the dataset are shown in Table 4.2

Since human tracking is not the focus of the current work, we obtain human tracks through the active learning tool from [26]. The dataset along with the human tracks, and role annotations would be made publicly available. [2]

4.5.2 Role Discovery Results

In our experiments, we evaluate the model by comparing results with human anno-tated roles in each video. Due to the weakly supervised nature of the problem, we do not have a direct mapping between role clusters and ground-truth role labels. To facilitate easy comparison with different baselines and two variations of our method, the role clusters obtained from a method are each mapped to one of the human defined roles, maximizing the total correct role assignments in an event. We present results on the two datasets from Sect. 4.5.1 and compare our full model against different baselines in Tables 4.3, 4.4. The tables show the total accuracy of role assignment in an event. Our method does not require role labels for training unlike [9, 15]. More-over, the method in [9] was developed for egocentric videos and cannot be trivially

---

[2] https://sites.google.com/site/eevignesh/socialroles

Table 4.3 Total role assignment accuracy for the YouTube dataset

Method	Birthday (%)	Wedding (%)	Award function (%)	Physical training (%)
Prior	29.32	20.17	62.97	65.93
k-means cluster	33.88	29.43	31.97	57.67
CRF with Ψ_u	38.25	39.22	69.31	76.69
CRF with Ψ_{up}	41.53	38.83	77.75	77.91
Our model—$\Psi_p^{\mathrm{Prox.}}$	43.72	36.41	79.54	**82.82**
Our model—$\Psi_p^{Spat.}$	43.72	39.32	79.80	77.91
Our full model	**44.81**	**42.72**	**83.12**	**82.82**

The best performance in each event is marked in bold font

Table 4.4 Total role assignment accuracy for the TRECVID dataset

Method	Birthday	Wedding
Prior	28.72	21.63
k-means cluster	29.88	34.19
CRF with Ψ_u	35.98	38.71
CRF with Ψ_{up}	42.07	41.94
Our model—$\Psi_p^{\mathrm{Prox.}}$	41.46	41.29
Our model—$\Psi_p^{Spat.}$	43.90	41.29
Our full model	**44.51**	**43.87**

The best performance in each event is marked in bold font

extended to our dataset. Hence, we compare with other methods which do not require supervision as explained below.

- prior: Simple baseline, where a random person in each video is assigned the reference role, and the true prior of secondary roles is used to assign roles to other people in the video.
- k-means: Simple experiment, where people are clustered using appearance and spatiotemporal features.
- CRF with Ψ_u: In this baseline, we derive results without use of pairwise interaction featues. The intention of this experiment is to judge the importance of pairwise interaciton features in our model.
- CRF with Ψ_{up}: In this experiment, we find the mean pairwise interaction features of a person p_i^v, with every other person in the video to form a unary interaction feature. This unary interaction feature is concatenated with the origianl unary feature forming a combined Unary-Interaction feature $\Psi_{\mathrm{up}}(p_i^v)$ as shown in Eq. (4.14), and we run experiments similar to $U\ only$ with these features. This baseline is denoted by $Comb.\ UP$. This experiment uses human–human interaction as a context feature. The purpose of this experiment is to show the distinction in explicitly modeling the interaction between individual roles versus using interaction as a simple context feature.

(a)

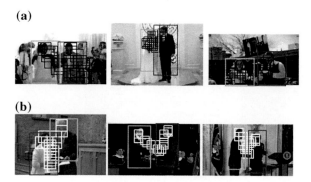

(b)

Fig. 4.4 Sample frames from videos are shown, where our full model identified the correct **a** "bride" (*green box*), "groom"(*red box*) roles in *wedding* and **b** "presenter" (*green box*), "recipient" (*red box*) roles in *award function*. The same hand–hand touch code is seen to be detected on different instances of the same role pair. The *black* and *white boxes* are the part detections from two different proxemic models for hand–hand touch

$$\Psi_{\text{up}}(p_i^v) = \left[\Psi_u(p_i^v), \frac{\sum\limits_{p_j^v \neq p_i^v} \Psi_p(p_i^v, p_j^v)}{|\mathbb{P}_v|} \right] \tag{4.14}$$

- Our model—$\Psi_p^{\text{Prox.}}$: We use only the spatial consistency term along with unary features and neglect the proxemic interaction term.
- Our model—Ψ_p^{ST}: We use only the proxemic interaction term along with unary features and neglect the spatiotemporal interaction term.

From results in Table 4.3, we notice that a CRF using Ψ_u outperforms naive k-means clustering, justifying the use of this representation with our unary features. Also, the use of interaction as a context feature in Ψ_{up} is seen to do better than the use of only unary features, in most events. This confirms our belief that, human interactions are informative for role recognition. In particular, we observe a considerable increase for the *award function* event, where the interaction between the "recipient" and "presenter" as seen in Fig. 4.4b would help distinguish the "presenter" from other people at the dais. Next, we observe that our full model shows significant improvement over CRF with Ψ_{up}. This demonstrates the value in explicitly modeling interaction between role pairs, instead of using interaction as a context feature. For instance, consider a wedding with similar interactions between a "bride-groom" pair, and a "bridesmaid–groomsman" pair. These interactions lead to the same interaction-context feature, for both the "bride" and the "bridesmaid." However, our full model would treat them differently, due to the difference in the other role participating in the interaction, leading to a richer description.

Our full model using the complete pairwise interaction feature Ψ_p performs better than the models only using $\Psi_p^{\text{Prox.}}$ or Ψ_p^{ST}, showing the gain from use of both the components. It is interesting to note the considerable drop in performance for

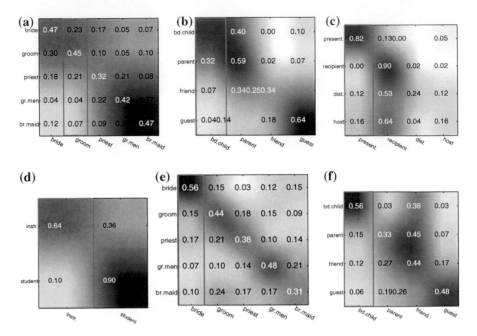

Fig. 4.5 Confusion matrices for different events are shown for the YouTube and TRECVID social roles dataset. The column corresponding to the reference role cluster chosen by our model is highlighted in each event. **a** wed. (YouTube), **b** b'day (YouTube), **c** award func., **d** p. train, **e** wed. (TRECVID), **f** b'day (TRECVID)

award function and *wedding* events, in the absence of $\Psi_p^{Prox.}$. We observed that the proxemic models corresponding to specific touch-codes fired consistently across different "bride-groom" and "presenter–recipient" pairs in *wedding* and *award functions* respectively, distinguishing them from other role pairs in the events. We illustrate this in Fig. 4.4.

To analyze the complete role assignment, we look at the confusion matrices in Fig. 4.5. The column corresponding to the reference role cluster chosen by our algorithm is highlighted in each matrix. The average purities of the reference role clusters are 0.65 and 0.56, in the YouTube and TRECVID datasets respectively. This demonstrates the ability of our model to isolate the reference role in each video. We observe that the model is able to cluster the roles better in the *wedding* event, as seen in Fig. 4.5a, e. This can be accounted to the strong interaction between the "bride" and "groom," separating them from the remaining roles. To study this interaction, we visualize the marginals of the spatial relationship of different roles with the reference role ("groom") cluster in the YouTube *wedding* dataset, in Fig. 4.6. The marginals capture the expected interaction, as explained in the figure. The confusion of "distributor" with the "recipient" in Fig. 4.5c, can be explained by the similar patterns of interaction between the "recipient" receiving the award from the "presenter," and the "distributor" handing out the award to the "presenter." "friends" are difficult to

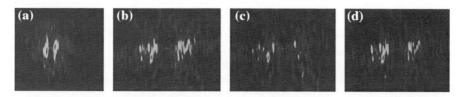

Fig. 4.6 Marginal of the position of a role relative to the reference ("groom"), estimated by our model is shown for YouTube *wedding* videos. The spatial positions in the image co-ordinates are normalized by width of the reference role bounding box, and binned. The groom's position is marked by a cross-hair. The "bride" is mostly close to the "groom." "groomsmen" and "bridesmaids" are distributed around the groom as expected. The uncertainty in recognizing the "priest" is reflected by a scattered distribution **a** bride, **b** priest, **c** bridesmaids, **d** Groomsmen

Fig. 4.7 Sample results from the YouTube social roles dataset is shown, where each row corresponds to an event. Boxes with *solid lines* indicate correct role assignments from our full model, while *dashed lines* represent faulty assignments. Different roles are indicated by the same color code as in Fig. 4.2. The ground truth role of a person is indicated by the *color* of the *dot* on the person. The last column shows typical failure cases for each event

distinguish from "guests" in the TRECVID *birthday* dataset, where we observed both roles to exhibit low interaction with the reference role.

Sample results from our full model are shown in Fig. 4.7 along with typical failure instances. Most failure cases involved less interaction among people, as seen in the last column of *birthday*, *wedding* and *physical training*.

In order to evaluate the latent reference role assignment in our model, we compare performances with a control setting which randomly chooses the reference role in each video. The average accuracy of role assignment over all events is seen to drop by 4.82 % for the YouTube social roles dataset with this choice of reference role, justifying the need to model it as a latent variable. In particular, we observe a large drop of 6.80 % for the *wedding* event, which has more role classes than the other events leading to increased randomness in the choice of reference role in each video.

4.6 Summary

In this chapter, we introduced the problem of recognizing social roles played by people in an event. Social roles are governed by human interactions, and form a fundamental component of human event description. We focused on a weakly supervised setting, where we were provided different videos belonging to an event class along with the human tracks, without training role labels. This weak supervision enables our method to automatically understand the relations between people, and discover the different roles associated with an event. It further reduces the human effort involved in observing long video footages to annotate the roles.

Social roles are event-specific and governed by the interactions between people in an event. We captured these interactions through pairwise interaction features in a CRF, along with person-specific social descriptors. We also developed tractable variational inference to simultaneously infer model weights, as well as role assignment to all people in the videos. Unlike naive clustering approaches like k-means, this allowed us to incorporate role-specific relations like "bride-groom" and "parent–child" interactions in the model. The method was evaluated on a novel YouTube social roles dataset with ground truth role annotations, and a subset of videos from the TRECVID-MED11 [1] event kits. We showed considerable performance improvement over different baseline models. As a next step, our approach can be extended to perform simultaneous event classification along with role discovery. It is also noted that our method is not robust to noisy and fragmented reference role tracking, due to the inherent assumption of one reference role per video. In the future, we wish to account for such noisy tracking.

Acknowledgments We thank A. Alahi, J. Krause, and K. Tang for helpful comments. This research is partially supported by the DARPA-Mind's Eye grant, and the IARPA-Aladdin grant.

References

1. Trecvid multimedia event detection track. http://www.nist.gov/itl/iad/mig/med11.cfm (2011)
2. Biddle, B.J.: Recent development in role theory. Ann. Rev. Sociol. **12**, 67–92 (1986)
3. Burgos-Artizzu, X., Dollar, P., Lin, D., Anderson, D., Perona, P.: Social behavior recognition in continuous videos. In: CVPR (2012)

4. Choi, W., Savarese, S.: A unified framework for multi-target tracking and collective activity recognition. In: ECCV (2012)
5. Cristani, M., Paggetti, G., Fossati, A., Bazzani, L., Tosato, D., Bue, A.D., Menegaz, G., Murino, V.: Social interaction discovery by statistical analysis of f-formations. In: BMVC (2011)
6. Ding, L., Yilmaz, A.: Learning relations among movie characters: a social network perspective. In: ECCV (2010)
7. Ding, L., Yilmaz, A.: Inferring social relations from visual concepts. In: ICCV (2011)
8. Direkolu, C., OConnor, N.: Team activity recognition in sports. In: ECCV (2012)
9. Fathi, A., Hoggins, J.K., Rehg, J.M.: Social interactions: a first person perspective. In: CVPR (2012)
10. Felzenszwalb, P., Girshick, R., McAllester, D., Ramanan, D.: Object detection with discriminatively trained part based models. IEEE Trans. Pattern Anal. Mach. Intell. **32**(9), 1627–1645 (2010)
11. Fu, Y., Hospedales, T., Xiang, T., Gong, S.: Attribute learning for understanding unstructured social activity, In: ECCV (2012)
12. Gallagher, A.C., Chen, T.: Understanding images of groups of people. In: CVPR (2009)
13. Kläser, A., Marszałek, M., Schmid, C.: A spatio-temporal descriptor based on 3d-gradients. In: BMVC (2008)
14. Klaser, A., Schmid, C., Liu, C.-L.: Action recognition by dense trajectories. In: CVPR (2011)
15. Lan, T., Sigal, L., Mori, G.: Social roles in hierarchical models for human activity recognition. In: CVPR (2012)
16. Lan, T., Wang, Y., Yang, W., Robinovitch, S., Mori, G.: Discriminative latent models for recognizing contextual group activities. IEEE Trans. Pattern Anal. Mach. Intell. **34**(8), 1549–1562 (2012)
17. Li, L.-J., Su, H., Xing, E.P., Fei-Fei, L.: Object bank: a high-level image representation for scene classification and semantic feature sparsification. In: NIPS (2010)
18. Li, R., Porfilio, P., Zickler, T.: Finding group interactions in social clutter. In: CVPR (2013)
19. Liu, D., Dong, C., Nocedal, J.: On the limited memory bfgs method for large scale optimization. Math. Program. **45**, 503–528 (1989)
20. Marin-Jimenez, M., Zisserman, A., Ferrari. V.: Heres looking at you, kid-detecting people looking at each other in videos. In: BMVC (2011)
21. Perez, A.P., Marszalek, M., Zisserman, A., Reid, I.: High five: recognising human interactions in tv shows. In: BMVC (2010)
22. Qin, Z., Shelton, C.R.: Improving multi-target tracking via social grouping. In: CVPR (2012)
23. Ramanathan, V., Yao, B., Fei-Fei, L.: Social role discover in human events. In: CVPR (2013)
24. Song, Z., Wang, M., Hua, X., Yan, S.: Predicting occupation via human clothing and contexts. In: ICCV (2011)
25. Stone, Z., Zickler, T., Darrell, T.: Toward large-scale face recognition using social network context. Proc. IEEE **98**(8), 1408 (2010)
26. Vondrick, C., Ramanan, D.: Video annotation and tracking with active learning. In: NIPS (2011)
27. Wang, G., Gallagher, A., Luo, J., Forsyth, D.: Seeing people in social context: recognizing people and social relationships. In: ECCV (2010)
28. Weng, C.-Y., Chu, W.-T., Rolenet, J-LWu: Movie analysis from the perspective of social networks. IEEE Trans. Multimedia **2**, 256–271 (2009)
29. Yang, Y., Baker, S., Kannan, A., Ramanan, D.: Recognizing proxemics in personal photos. In: CVPR (2012)
30. Yu, T., Lim, S.-N., Patwardhan, K., Krahnstoever, N.: Monitoring, recognizing and discovering social networks. In: CVPR (2009)
31. Zhu, J., Xing, E.P.: Conditional topic random fields. In: ICML (2010)
32. Zhu, X., Ramanan, D.: Face detection, pose estimation, and landmark localization in the wild. In: CVPR (2012)

Chapter 5
Integrating Randomization and Discrimination for Classifying Human-Object Interaction Activities

Aditya Khosla, Bangpeng Yao and Li Fei-Fei

5.1 Introduction

Psychologists have shown that the ability of humans to perform basic-level categorization (e.g., cars vs. dogs; kitchen vs. highway) develops well before their ability to perform subordinate-level categorization, or fine-grained visual categorization (e.g., distinguishing dog breeds such as Golden retrievers vs. Labradors) [18]. It is interesting to observe that computer vision research has followed a similar trajectory. Basic-level object and scene recognition has seen great progress [15, 21, 26, 31] while fine-grained categorization has received little attention. Unlike basic-level recognition, even humans might have difficulty with some of the fine-grained categorization [32]. Thus, an automated visual system for this task could be valuable in many applications.

Action recognition in still images can be regarded as a fine-grained classification problem [17] as the action classes only differ by human pose or type of human–object interactions. Unlike traditional object or scene recognition problems where different classes can be distinguished by different parts or coarse spatial layout [15, 16, 21], more detailed visual distinctions need to be explored for fine-grained image classification. The bounding boxes in Fig. 5.1 demarcate the distinguishing characteristics between closely related bird species, or different musical instruments or human poses

---

An early version of this chapter was presented in Yao et al. [37], and the code is available at http://vision.stanford.edu/discrim_rf/.

A. Khosla (✉)
MIT, 32 Vasar St, Cambridge, MA 2139, USA
e-mail: khosla@csail.mit.edu

B. Yao · L. Fei-Fei
Stanford University, Stanford, CA, USA
e-mail: bangpeng@cs.stanford.edu

L. Fei-Fei
e-mail: feifeili@cs.stanford.edu

Playing Clarinet   Holding Clarinet   Playing Recorder

California Gull   Glaucous-winged Gull   Heerman Gull

Fig. 5.1 Human action recognition (*top row*) is a fine-grained image classification problem, where the human body dominates the image. It is similar to the subordinate object classification problem (*bottom row*). The *red* and *yellow* bounding boxes indicate discriminative image patches for both tasks (manually drawn for illustration). The goal of our algorithm is to discover such discriminative image patches automatically

that differentiate the different playing activities. Models and algorithms designed for basic-level object or image categorization tasks are often unprepared to capture such subtle differences among the fine-grained visual classes. In this chapter, we approach this problem from the perspective of finding a large number of *image patches* with arbitrary shapes, sizes, or locations, as well as associations between pairs of patches that carry discriminative image statistics [9, 33] (Sect. 5.3). However, this approach poses a fundamental challenge: without any feature selection, even a modestly sized image will yield millions or billions of image patches. Furthermore, these patches are highly correlated because many of them overlap significantly. To address these issues, we propose the use of *randomization* that considers a random subset of features at a time.

Specifically, we propose a *random forest with discriminative decision trees* algorithm to discover image patches and pairs of patches that are highly discriminative for fine-grained categorization tasks. Unlike conventional decision trees [2, 4, 8], our algorithm uses strong classifiers at each node and combines information at different depths of the tree to effectively mine a very dense sampling space. Our method significantly improves the strength of the decision trees in the random forest while still maintaining low correlation between the trees. This allows our method to achieve low generalization error according to the theory of random forest [4].

Besides action recognition in still images [11, 12, 33], we evaluate our method on subordinate categorization of closely related animal species [32]. We show that our method achieves state-of-the-art results. Furthermore, our method identifies semantically meaningful image regions[1] that closely match human intuition. Additionally, our method tends to automatically generate a coarse-to-fine structure of discriminative image patches, which parallels the human visual system [5].

The remaining part of this chapter is organized as follows: Section 5.2 discusses related work. Section 5.3 describes our dense feature space and Sect. 5.4 describes our algorithm for mining this space. Experimental results are discussed in Sects. 5.5, and 5.6 summarizes this chapter.

5.2 Related Work

Image classification has been studied for many years. Most of the existing work focuses on basic-level categorization such as objects [2, 14, 15] or scenes [13, 21, 26]. In this chapter we focus on two tasks of fine-grained image classification: (1) identifying human–object interaction activities in still images [34–36, 39], and subordinate-level categorization of animal species [3, 17, 20, 38], which requires an approach that captures the fine and detailed information in images.

In this chapter, we explore a dense feature representation to distinguish fine-grained image classes. "Grouplet" features [33] have shown the advantage of dense features in classifying human activities. Instead of using the generative local features as in Grouplet, here we consider a richer feature space in a discriminative setting where both local and global visual information are fused together. Inspired by [9, 33], our approach also considers pairwise interactions between image regions.

We use a random forest framework to identify discriminative image regions. Random forests have been used successfully in many vision tasks such as object detection [2], segmentation [27], and codebook learning [24]. Inspired from Tu [28], we combine discriminative training and randomization to obtain an effective classifier with good generalizability. Our method differs from Tu [28] in that for each tree node, we train an SVM classifier from one of the randomly sampled image regions, instead of using AdaBoost to combine weak features from a fixed set of regions. This allows us to explore an extremely large feature set efficiently.

A classical image classification framework [31] is *Feature Extraction* → *Coding* → *Pooling* → *Concatenating*. *Feature extraction* [23] and better *coding* and *pooling* methods [31] have been extensively studied for object recognition. In this work, we use discriminative feature mining and randomization to propose a new feature *concatenating* approach, and demonstrate its effectiveness on fine-grained image categorization tasks.

---

[1] We use the terms "patches" and "regions" interchangeably throughout this chapter.

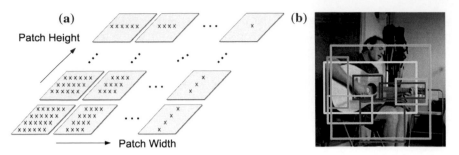

Fig. 5.2 Illustration of the proposed dense sampling space. **a** We densely sample rectangular image patches with varying widths and heights. The regions are closely located and have significant overlaps. The *red* × denote the centers of the patches, and the *arrows* indicate the increment of the patch width or height. **b** Illustration of some image patches that may be discriminative for "playing-guitar." All those patches can be sampled from our dense sampling space

5.3 Dense Sampling Space

Our algorithm aims to identify fine image statistics that are useful for fine-grained categorization. For example, in order to classify whether a human is playing a guitar or holding a guitar without playing it, we want to use the image patches below the human face that are closely related to the human–guitar interaction (Fig. 5.2b). An algorithm that can reliably locate such regions is expected to achieve high classification accuracy. We achieve this goal by searching over rectangular image patches of arbitrary width, height, and image location. We refer to this extensive set of image regions as the *dense sampling space*, as shown in Fig. 5.2a. This figure has been simplified for visual clarity, and the actual density of regions considered in our algorithm is significantly higher. We note that the regions considered by spatial pyramid matching (SPM) [21] is a very small subset lying along the diagonal of the height–width plane that we consider. Further, to capture more discriminative distinctions, we also consider interactions between pairs of arbitrary patches. The pairwise interactions are modeled by applying concatenation, absolute of difference, or intersection between the feature representations of two image patches.

However, the dense sampling space is very huge. Sampling image patches of size 50×50 in a 400×400 image every four pixels leads to thousands of patches. This increases many folds when considering regions with arbitrary widths and heights. Further considering pairwise interactions of image patches will effectively lead to trillions of features for each image. In addition, there is much noise and redundancy in this feature set. On the one hand, many image patches are not discriminative for distinguishing different image classes. On the other hand, the image patches are highly overlapped in the dense sampling space, which introduces significant redundancy among these features. Therefore, it is challenging to explore this high-dimensional, noisy, and redundant feature space. In this work, we address this issue using randomization.

---

foreach *tree t* **do**
  - Sample a random set of training examples $\mathscr{D}$;
  - SplitNode($\mathscr{D}$);
  if *needs to split* **then**
    i. Randomly sample the candidate (pairs of) image regions (Sect. 5.4.2);
    ii. Select the best region to split $\mathscr{D}$ into two sets $\mathscr{D}_1$ and $\mathscr{D}_2$ (Sect. 5.4.3);
    iii. Go to SplitNode($\mathscr{D}_1$) and SplitNode($\mathscr{D}_2$).
  else
    Return $P_t(c)$ for the current leaf node.
  end
end

Algorithm 1: Overview of the process of growing decision trees in the random forest framework.

5.4 Discriminative Random Forest

In order to explore the dense sampling feature space for fine-grained visual categorization, we combine two concepts: (1) *Discriminative training* to extract the information in the image patches *effectively*; (2) *Randomization* to explore the dense feature space *efficiently*. Specifically, we adopt a random forest [4] framework where each tree node is a discriminative classifier that is trained on one or a pair of image patches. In our setting, the discriminative training and randomization can benefit from each other. We summarize the advantages of our method below:

- The random forest framework allows us to consider a subset of the image regions at a time, which allows us to explore the dense sampling space efficiently in a principled way.
- Random forest selects a best image patch in each node, and therefore it can remove the noise-prone image patches and reduce redundancy in the feature set.
- By using discriminative classifiers to train the tree nodes, our random forest has much stronger decision trees. Further, because of the large number of possible image regions, it is likely that different decision trees will use different image regions, which reduces the correlation between decision trees. Therefore, our method is likely to achieve low generalization error (Sect. 5.4.4) compared with the traditional random forest [4] which uses weak classifiers in the tree nodes.

An overview of the random forest framework we use is shown in Algorithm 1. In the following sections, we first describe this framework (Sect. 5.4.1). Then we elaborate on our feature sampling (Sect. 5.4.2) and split learning (Sect. 5.4.3) strategies in detail, and describe the generalization theory [4] of random forest which guarantees the effectiveness of our algorithm (Sect. 5.4.4).

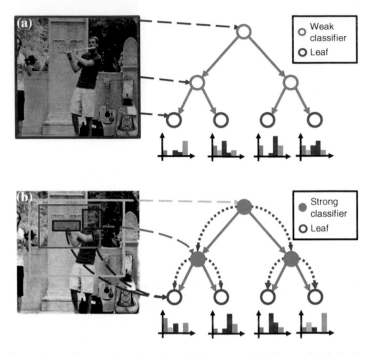

Fig. 5.3 Comparison of conventional random decision trees (**a**) with our discriminative decision trees (**b**). *Solid blue arrows* show binary splits of the data. *Dotted lines* from the shaded image regions indicate the region used at each node. Conventional decision trees use information from the entire image at each node, which encodes no spatial or structural information, while our decision trees sample single or multiple image regions from the dense sampling space (Fig. 5.2a). The histograms below the leaf nodes illustrate the posterior probability distribution $P_{t,l}(c)$ (Sect. 5.4.1). In **b**, *dotted red arrows* between nodes show our nested tree structure that allows information to flow in a *top–down* manner. Our approach uses strong classifiers in each node (Sect. 5.4.3), while the conventional method uses weak classifiers

5.4.1 The Random Forest Framework

Random forest is a multiclass classifier consisting of an ensemble of decision trees where each tree is constructed via some randomization. As shown in Fig. 5.3a, the leaf nodes of each tree encode a distribution over the image classes. All internal nodes contain a binary test that splits the data and sends the splits to its children nodes. The splitting is stopped when a leaf node is encountered. An image is classified by descending each tree and combining the leaf distributions from all the trees. This method allows the flexibility to explore a large feature space effectively because it only considers a subset of features in every tree node.

Each tree returns the posterior probability of an example belonging to the given classes. The posterior probability of a particular class at each leaf node is learned as the proportion of the training images belonging to that class at the given leaf node. The posterior probability of class c at leaf l of tree t is denoted as $P_{t,l}(c)$. Thus, a test image can be classified by averaging the posterior probability from the leaf node of each tree:

$$c^* = \arg\max_c \frac{1}{T} \sum_{t=1}^{T} P_{t,l_t}(c), \tag{5.1}$$

where c^* is the predicted class label, T is the total number of trees, and l_t is the leaf node that the image falls into.

In the following sections, we describe the process of obtaining $P_{t,l}(c)$ using our algorithm. Readers can refer to previous works [2, 4, 27] for more details of the conventional decision tree learning procedure.

5.4.2 Sampling the Dense Feature Space

As shown in Fig. 5.3b, each internal node in our decision tree corresponds to a single or a pair of rectangular image regions that are sampled from the dense sampling space (Sect. 5.3), where the regions can have many possible widths, heights, and image locations. In order to sample a candidate image region, we first normalize all images to unit width and height, and then randomly sample (x_1, y_1) and (x_2, y_2) from a uniform distribution $U([0, 1])$. These coordinates specify two diagonally opposite vertices of a rectangular region. Such regions could correspond to small areas of the image (e.g., the purple bounding boxes in Fig. 5.3b) or even the complete image. This allows our method to capture both global and local information in the image.

In our approach, each sampled image region is represented by a histogram of visual descriptors. For a pair of regions, the feature representation is formed by applying histogram operations (e.g., concatenation, intersection, etc.) to the histograms obtained from both regions. Furthermore, the features are augmented with the decision value $\mathbf{w}^T\mathbf{f}$ (described in Sect. 5.4.3) of this image from its parent node (indicated by the dashed red lines in Fig. 5.3b). Therefore, our feature representation combines the information of all upstream tree nodes that the corresponding image has descended from. We refer to this idea as "nesting." Using feature sampling and nesting, we obtain a candidate set of features, $\mathbf{f} \in \mathbb{R}^n$, corresponding to a candidate image region of the current node.

Implementation details. Our method is flexible to use many different visual descriptors. In this work, we densely extract SIFT [23] descriptors on each image with a spacing of four pixels. The scales of the grids to extract descriptors are 8, 12, 16, 24, and 30. Using k-means clustering, we construct a vocabulary of codewords.[2] Then, we use Locality-constrained Linear Coding [31] to assign the descriptors to codewords. A bag-of-words histogram representation is used if the area of the patch is smaller than 0.2, while a 2-level or 3-level spatial pyramid is used if the area is

---

[2] A dictionary size of 1024, 256, 256 is used for PASCAL action [11, 12], PPMI [33], and Caltech-UCSD Birds [32] datasets respectively.

between 0.2 and 0.8 or larger than 0.8, respectively. Note that all parameter here are empirically chose. Using other similar parameters will lead to very similar results.

During sampling (step i of Algorithm 1), we consider four settings of image patches: a single image patch and three types of pairwise interactions (concatenation, intersection, and absolute of difference of the two histograms). We sample 25 and 50 image regions (or pairs of regions) in the root node and the first level nodes, respectively, and sample 100 regions (or pairs of regions) in all other nodes. Sampling a smaller number of image patches in the root can reduce the correlation between the resulting trees.

5.4.3 Learning the Splits

In this section, we describe the process of learning the binary splits of the data using SVM (step ii in Algorithm 1). This is achieved in two steps: (1) Randomly assigning all examples from each class to a binary label; (2) Using SVM to learn a binary split of the data.

Assume that we have C classes of images at a given node. We uniformly sample C binary variables, $\mathbf{b}$, and assign all examples of a particular class c_i a class label of b_i. As each node performs a binary split of the data, this allows us to learn a simple binary SVM at each node. This improves the scalability of our method to a large number of classes and results in well-balanced trees. Using the feature representation $\mathbf{f}$ of an image region (or pairs of regions) as described in Sect. 5.4.2, we find a binary split of the data:

$$\begin{cases} \mathbf{w}^{\mathrm{T}}\mathbf{f} \leq 0, \text{ go to left child} \\ \text{otherwise, go to right child} \end{cases}$$

where $\mathbf{w}$ is the set of weights learned from a linear SVM.

We evaluate each binary split that corresponds to an image region or pairs of regions with the information gain criteria [2], which is computed from the complete training images that fall at the current tree node. The splits that maximize the information gain are selected and the splitting process (step iii in Algorithm 1) is repeated with the new splits of the data. The tree splitting stops if a pre-specified maximum tree depth has been reached, or the information gain of the current node is larger than a threshold, or the number of samples in the current node is small.

5.4.4 Generalization Error of Random Forests

In Breiman [4], it has been shown that an upper bound for the generalization error of a random forest is given by

$$\rho(1 - s^2)/s^2, \tag{5.2}$$

where s is the strength of the decision trees in the forest, and ρ is the correlation between the trees. Therefore, the generalization error of a random forest can be reduced by making the decision trees stronger or reducing the correlation between the trees.

In our approach, we learn discriminative SVM classifiers for the tree nodes. Therefore, compared to the traditional random forests where the tree nodes are weak classifiers of randomly generated feature weights [2], our decision trees are much stronger. Furthermore, since we are considering an extremely dense feature space, each decision tree only considers a relatively small subset of image patches. This means there is little correlation between the trees. Therefore, our random forest with discriminative decision trees algorithm can achieve very good performance on fine-grained image classification, where exploring fine image statistics discriminatively is important. In Sect. 5.5.5, we show the strength and correlation of different settings of random forests with respect to the number of decision trees, which justifies the above arguments. Please refer to Breiman [4] for details about how to compute the strength and correlation values for a random forest.

5.5 Experiments

In this section, we first evaluate our algorithm on two fine-grained image datasets: actions of people-playing-musical-instrument (PPMI) [33] (Sect. 5.5.1) and a subordinate object categorization dataset of 200 bird species [32] (Sect. 5.5.2). Experimental results show that our algorithm outperforms state-of-the-art methods on these datasets. Further, we use the proposed method to participate the action classification competition of the PASCAL VOC challenge, and obtain the winning award in both 2011 [11] and 2012 [12]. Detailed results and analysis are shown in Sects. 5.5.3 and 5.5.4. Finally, we evaluate the strength and correlation of the decision trees in our method, and compare the result with the other settings of random forests to show why our method can lead to better classification performance (Sect. 5.5.5).

5.5.1 People-Playing-Musical-Instruments

The people-playing-musical-instrument (PPMI) dataset is introduced in Yao and Fei-Fei [33]. This dataset puts emphasis on understanding subtle interactions between humans and objects. Here we use a full version of the dataset which contains 12 musical instruments; for each instrument there are images of people playing the instrument and holding the instrument but not playing it. We evaluate the performance of our method with 100 decision trees on the 24-class classification problem. We compare our method with many previous results,[3] including bag of words, grouplet

---

[3] The baseline results are available from the dataset website: http://ai.stanford.edu/~bangpeng/ppmi

Table 5.1 Mean Average Precision (% mAP) on the 24-class classification problem of the PPMI dataset

Method	BoW	Grouplet [33]	SPM [21]	LLC [31]	Ours
mAP (%)	22.7	36.7	39.1	41.8	**47.0**

The best result is highlighted with bold fonts

Table 5.2 Comparison of mean Average Precision (% mAP) of the results obtained by different methods on the PPMI binary classification tasks of people playing and holding different musical instruments

Instrument	BoW	Grouplet [33]	SPM [21]	LLC [31]	Ours
Bassoon	73.6	78.5	84.6	85.0	**86.2**
Erhu	82.2	87.6	88.0	89.5	**89.8**
Flute	86.3	95.7	95.3	97.3	**98.6**
French horn	79.0	84.0	93.2	93.6	**97.3**
Guitar	85.1	87.7	**93.7**	92.4	93.0
Saxophone	84.4	87.7	89.5	88.2	**92.4**
Violin	80.6	93.0	93.4	**96.3**	95.7
Trumpet	69.3	76.3	82.5	86.7	**90.0**
Cello	77.3	84.6	85.7	82.3	**86.7**
Clarinet	70.5	82.3	82.7	84.8	**90.4**
Harp	75.0	87.1	92.1	**93.9**	92.8
Recorder	73.0	76.5	78.0	79.1	**92.8**
Average	78.0	85.1	88.2	89.2	**92.1**

Each column shows the results obtained from one method. The best results are highlighted with bold fonts

[33], SPM [21], locality-constrained linear coding (LLC) [31]. The grouplet method uses one SIFT scale, while all the other methods use multiple SIFT scales described in Sect. 5.4.2. Table 5.1 shows that we significantly outperform the a various of previous approaches.

Table 5.2 shows the result of our method on the 12 binary classification tasks where each task involves distinguishing the activities of playing and not playing for the same instrument. Despite a high baseline of 89.2 % mAP, our method outperforms by 2.9 % to achieve a result of 92.1 % overall. We also perform better than the grouplet approach [33] by 7 %, mainly because the random forest approach is more expressive. While each grouplet is encoded by a single visual codeword, each node of the decision trees here corresponds to an SVM classifier. Furthermore, we outperform the baseline methods on 9 of the 12 binary classification tasks. In Fig. 5.4, we visualize the heat map of the features learned for this task. We observe that they show semantically meaningful locations of where we would expect the discriminative regions of people playing different instruments to occur. For example, for flute, the region around the face provides important information while for guitar, the region to the left of the torso provides more discriminative information. It is interesting to note that despite the randomization and the algorithm having no prior information, it is able to locate the region of interest reliably.

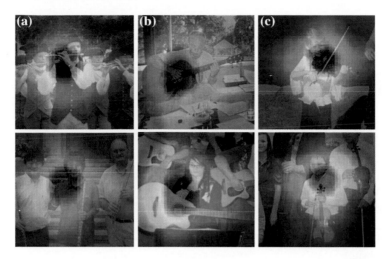

Fig. 5.4 **a** Heat map of the dominant regions of interest selected by our method for playing flute on images of playing flute (*top row*) and holding a flute without playing it (*bottom row*). **b, c** shows similar images for guitar and violin, respectively. The heat maps are obtained by aggregating image regions of all the tree nodes in the random forest weighted by the probability of the corresponding class. *Red* indicates high frequency and *blue* indicates low frequency

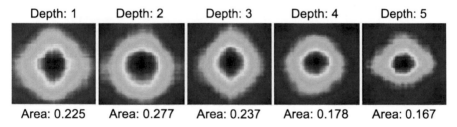

Fig. 5.5 Heat map for "playing trumpet" class with the weighted average area of selected image regions for each tree depth. Please refer to Fig. 5.4 for how the heat maps are obtained

Furthermore, we also demonstrate that the method learns a coarse-to-fine region of interest for identification. This is similar to the human visual system which is believed to analyze raw input in order from low to high spatial frequencies or from large global shapes to smaller local ones [5]. Figure 5.5 shows the heat map of the area selected by our classifier as we consider different depths of the decision tree. We observe that our random forest follows a similar coarse-to-fine structure. The average area of the patches selected reduces as the tree depth increases. This shows that the classifier first starts with more global features or high frequency features to discriminate between multiple classes, and finally zeros in on the specific discriminative regions for some particular classes.

Table 5.3 Comparison of the mean classification accuracy of our method and the baseline results on the Caltech-UCSD Birds 200 dataset

Method	MKL [3]	LLC [31]	Ours
Accuracy	19.0%	18.0%	**19.2%**

The best performance is indicated with bold fonts

5.5.2 Caltech-UCSD Birds 200

The Caltech-UCSD Birds (CUB-200) dataset contains 6,033 annotated images of 200 different bird species [32]. This dataset has been designed for subordinate image categorization. It is a very challenging dataset as the different species are very closely related and have similar shape/color. There are around 30 images per class with 15 for training and the remaining for testing. The test-train splits are fixed (provided on their website).

The images are cropped to the provided bounding box annotations. These regions are resized such that the smaller image dimension is 150 pixels. As color provides important discriminative information, we extract C-SIFT descriptors [29] in the same way described in Sect. 5.4.2. We use 300 decision trees in our random forest. Table 5.3 compares the performance of our algorithm against the LLC baseline and the state-of-the-art result (multiple kernel learning (MKL) [3]) on this dataset. Our method outperforms LLC and achieves comparable performance with the MKL approach. We note that Branson et al. [3] uses multiple features, e.g., geometric blur, gray/color SIFT, full image color histograms, etc. It is expected that including these features can further improve the performance of our method. Furthermore, we show in Fig. 5.6 that our method is able to capture the intraclass pose variations by focusing on different image regions for different images.

5.5.3 PASCAL VOC 2011 Action Classification

The recent PASCAL VOC challenges incorporated the task of recognizing actions in still images. The images describe ten common human activities: "Jumping," "Phoning," "Playing a musical instrument," "Reading," "Riding a bicycle or motorcycle," "Riding a horse," "Running," "Taking a photograph," "Using a computer," and "Walking." Each person that we need to classify is indicated by a bounding box and is annotated with one of the nine actions they are performing. There are also humans performing actions that do not belong to any of the ten aforementioned categories. These actions are all labeled as "Other."

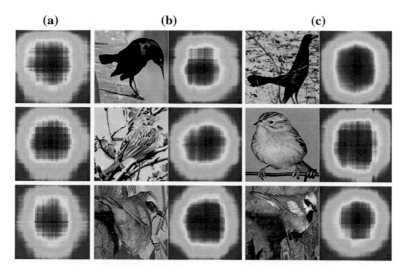

Fig. 5.6 Each row represents visualizations for a single class of birds (from *top* to *bottom*): boat tailed grackle, brewer sparrow, and golden winged warbler. For each class, we visualize: **a** Heat map for the given bird as described in Fig. 5.4; **b, c** Two example images of the corresponding bird and the distribution of image patches selected for the specific image

We participated the competition using the method proposed in this chapter, and won the winning award in both 2011 [11][4] and 2012 [12].[5] We introduce the details of our results in the 2011 challenge [11] in the rest of this subsection. Section 5.5.4 will cover our results in the 2012 challenge [12].

There are around 2,500 training/validation images and a similar number of testing images in the 2011 dataset. As in Delaitre et al. [7], we obtain a foreground image for each person by extending the bounding box of the person to contain 1.5× the original size of the bounding box, and resizing it such that the larger dimension is 300 pixels. We also resize the original image accordingly. Therefore for each person, we have a "person image" as well as a "background image." We only sample regions from the foreground and concatenate the features with a two-level spatial pyramid of the background. We use 100 decision trees in our random forest.

Classification results measured by mean Average Precision (mAP) are shown in Table 5.4. Our method achieves the best result on six out of the ten actions. Note that we achieved this accuracy based on only grayscale SIFT descriptors, without using any other features or contextual information like object detectors.

---

[4] A summary of the results in 2011 PASCAL challenge is in http://pascallin.ecs.soton.ac.uk/challenges/VOC/voc2011/workshop/index.html http://pascallin.ecs.soton.ac.uk/challenges/VOC/voc2011/workshop/index.html.

[5] A summary of the results in 2012 PASCAL challenge is in http://pascallin.ecs.soton.ac.uk/challenges/VOC/voc2012/workshop/index.html http://pascallin.ecs.soton.ac.uk/challenges/VOC/voc2012/workshop/index.html.

Table 5.4 Comparison of the mean Average Precision of our method and the other approaches in the action classification competition of PASCAL VOC 2011

Action	CAENLEAR DSAL (%)	CAENLEAR HOBJ_DSAL (%)	NUDT CONTEXT (%)	NUDT SEMANTIC (%)	Ours (%)
Jumping	62.1	**71.6**	65.9	66.3	66.0
Phoning	39.7	**50.7**	41.5	41.3	41.0
Playing instrument	60.5	**77.5**	57.4	53.9	60.0
Reading	33.6	37.8	34.7	35.2	**41.5**
Riding bike	80.8	86.5	88.8	88.8	**90.0**
Riding horse	83.6	89.5	90.2	90.0	**92.1**
Running	80.3	83.8	**87.9**	87.6	86.6
Taking photo	23.2	25.1	25.7	25.5	**28.8**
Using computer	53.4	58.9	54.5	53.7	**62.0**
Walking	50.2	59.2	59.5	58.2	**65.9**

Each column shows the result from one method. The best results are highlighted with bold fonts. We omitted the results of MISSOURI_SSLMF and WVU_SVM-PHOW, which did not outperform on any class, due to space limitations

Figure 5.7 shows the frequency of an image patch being selected by our method. For each activity, the figure is obtained by considering the features selected in the tree nodes weighted by the proportion of samples of this activity in this node. From the results, we can clearly see the difference of distributions for different activities. For example, the image patches corresponding to human–object interactions are usually highlighted, such as the patches of bikes and books. We can also see that the image patches corresponding to background are not frequently selected. This demonstrates our algorithm's ability to deal with background clutter.

5.5.4 PASCAL VOC 2012 Action Classification

The action classification competition of the 2012 PASCAL VOC challenge [12] contains more than 5,000 training/validation images and a similar number of testing images, which is an increase of around 90 % in size over VOC 2011. We use our proposed method with two improvements: (1) combining multiple features, and (2) greedy tree selection. We describe these in greater detail below. The results are shown in Table 5.5. In 2012 we had only one competitor (DPM_RF_SVM), and our method outperformed this approach on eight of the ten action classes. Further, comparing "Ours 2012" with "Ours 2011," we observe that combining multiple features and using a tree selection approach[6] improves the performance by 6 % mAP.

---

[6] These approaches were specifically developed for the 2012 PASCAL VOC challenge and have not been tested on other datasets but we expect similar performance improvements on them.

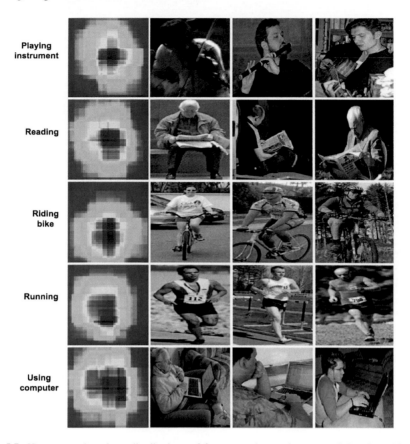

Fig. 5.7 Heat maps that show distributions of frequency that an image patch is selected in our method. Please refer to Fig. 5.4 for an explanation on how the heat maps are obtained

Table 5.5 Comparison of the mean Average Precision of our method and the other approaches in the action classification competition of PASCAL VOC 2012

Action	DPM_RF_SVM (%)	Ours 2011 (%)	Ours 2012 (%)
Jumping	73.8	71.1	**75.7**
Phoning	**45.0**	41.2	44.8
Playing instrument	62.8	61.9	**66.6**
Reading	41.4	39.3	**44.4**
Riding bike	93.0	92.4	**93.2**
Riding horse	93.4	92.5	**94.2**
Running	**87.8**	86.1	87.6
Taking photo	35.0	31.3	**38.4**
Using computer	64.7	60.4	**70.6**
Walking	73.5	68.9	**75.6**

"Ours 2011" indicates our approach used for the 2011 challenge. The best results are highlighted with bold fonts

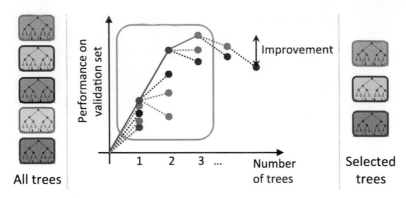

Fig. 5.8 Illustration of the process of greedy tree selection described in Sect. 5.5.4. *Left* Initially, we start with all the independently trained trees. *Middle* Then, we measure their performance on the validation data, one at a time. We select the tree with the highest validation performance in the first step (*blue*), and then choose from the remaining trees in the second step and so on. Overall, the improvement obtained by tree selection is indicated in the figure. *Right* Selected trees that maximize validation performance

Combining multiple features: Besides the SIFT image descriptor [23] used in the 2011 challenge, we also consider four other descriptors: HOG2x2 [6], color naming [30], local binary pattern [25], and object bank [22]. These features are extracted in a similar manner to [19]. For HOG2x2 and color naming features, we use a dictionary size of 1024 and 256 respectively. For object bank features, we train the deformable parts-based model (DPM) [15] on the 20 object categories in PASCAL VOC. We build decision trees for each feature independently. Then, we train a linear SVM on the class histograms obtained using the different features to obtain the final output.

Greedy tree selection: Figure 5.8 illustrates our algorithm. We use training images to train N decision trees independently, and then select the best subset of decision trees based on the validation performance in a greedy manner. We build the forest from decision trees in a sequential manner: first, we evaluate the performance of all individual decision trees on held-out validation data. Then, we select the tree that maximizes the validation performance. This results in a forest with one decision tree. We then evaluate the validation performance when we add one more tree from the remaining set of $N - 1$ trees and pick the tree that maximizes performance. We repeat this process for N trees, and select the best subset as the first $S \leq N$ trees that maximize the validation performance ($S = 3$ in Fig. 5.8). A greedy method (or another approximation) is required as there are too many possible subsets of trees (2^N) to enumerate exhaustively. The idea of tree selection has also been explored in prior works [1].

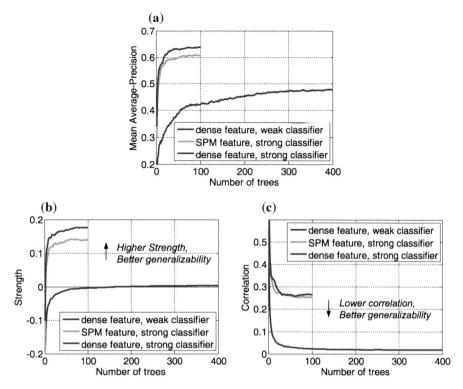

Fig. 5.9 Comparison of different random forest settings. **a** We compare the classification performance (mAP) obtained by our method dense feature, strong classifier with two control settings. Please refer to Sect. 5.5.5 for details of these settings. **b, c** We also compare the strength of the decision trees learned by these approaches and correlation between these trees (Sect. 5.4.4), which are highly related to the generalization error of random forests

5.5.5 Strength and Correlation of Decision Trees

We compare our method against two control settings of random forests on the PASCAL action dataset. Here we use the PASCAL VOC 2010 dataset [10] where there are fewer images than that on 2011 to make our experiments easier to conduct.

- *Dense feature, weak classifier*: For each image region or pairs of regions sampled from our dense sampling space, replace the SVM classifier in our method with a weak classifier as in the conventional decision tree learning approach [4, 8], i.e., randomly generating 100 sets of feature weights and select the best one.
- *SPM feature, strong classifier*: Use SVM classifiers to split the tree nodes as in our method, but the image regions are limited to that from a 4-level spatial pyramid.

Note that all other settings of the above two approaches remain unchanged as compared to our method (as described in Sect. 5.4). Figure 5.9 shows that on this dataset, a set of strong classifiers with relatively high correlation can lead to better performance than a set of weak classifiers with low correlation. We can see that the performance of random forests can be significantly improved by using strong classifiers in the nodes of decision trees. Compared to the random forests that only sample spatial pyramid regions, using the dense sampling space obtains stronger trees without significantly increasing the correlation between different trees, thereby improving the classification performance. Furthermore, the performance of the random forests using discriminative node classifiers converges with a small number of decision trees, indicating that our method is more efficient than the conventional random forest approach. In our experiment, the two settings and our method need a similar amount of time to train a single decision tree.

Additionally, we show the effectiveness of random binary assignment of class labels (Sect. 5.4.3) when we train classifiers for each tree node. Here we ignore this step and train a one-versus-all multiclass SVM for each sampled image region or pairs of regions. In this case C sets of weights are obtained when there are C classes of images at the current node. The best set of weights is selected using information gain as before. This setting leads to deeper and significantly unbalanced trees, and the performance decreases to 58.1 % with 100 trees. Furthermore, it is highly inefficient as it does not scale well with increasing number of classes.

5.6 Summary

In this chapter, we propose a random forest with discriminative decision trees algorithm to explore a dense sampling space for fine-grained image categorization. Experimental results on subordinate classification and activity classification show that our method achieves state-of-the-art performance and discovers much semantically meaningful information. One direction for future work is to extend the method to allow for more flexible regions where their location can vary from image to image. Furthermore, it would be interesting to apply other classifiers with analytical solutions such as Linear Discriminant Analysis to speed up the training procedure.

Acknowledgments L.F-F. is partially supported by an NSF CAREER grant (IIS-0845230), an ONR MURI grant, the DARPA VIRAT program and the DARPA Mind's Eye program. B.Y. is partially supported by the SAP Stanford Graduate Fellowship, and the Microsoft Research Fellowship. A.K. is supported by the Facebook Fellowship.

References

1. Bernard, S., Heutte, L., Adam, S.: On the selection of decision trees in random forests. In: IEEE International Joint Conference on Neural Networks, IJCNN, pp. 302–307 (2009)
2. Bosch, A., Zisserman, A., Munoz, X.: Image classification using random forests and ferns. In: Proceedings of the IEEE International Conference on Computer Vision (ICCV) (2007)

3. Branson, S., Wah, C., Babenko, B., Schroff, F., Welinder, P., Perona, P., Belongie, S.: Visual recognition with humans in the loop. In: Proceedings of the European Conference on Computer Vision (ECCV) (2010)
4. Breiman, L.: Random forests. Mach. Learn. **45**, 5–32 (2001)
5. Collin, C.A., McMullen, P.A.: Subordinate-level categorization relies on high spatial frequencies to a greater degree than basic-level categorization. Percept. Psychophys. **67**(2), 354–364 (2005)
6. Dalal, N., Triggs, B.: Histograms of oriented gradients for human detection. In: Proceedings of the IEEE Conference on Computer Vision and Pattern Recognition (CVPR) (2005)
7. Delaitre, V., Laptev, I., Sivic, J.: Recognizing human actions in still images: a study of bag-of-features and part-based representations. In: Proceedings of the British Machine Vision Conference (BMVC) (2010)
8. Dietterich, T.G.: An experimental comparison of three methods for constructing ensembles of decision trees: bagging, boosting, and randomization. Mach. Learn. **40**, 139–157 (2000)
9. Duan, G., Huang, C., Ai, H., Lao, S.: Boosting associated pairing comparison features for pedestrian detection.In: Proceedings of the Workshop on Visual Surveillance(2009)
10. Everingham, M., Van Gool, L., Williams, C., Winn, J., Zisserman, A.: The PASCAL Visual Object Classes Challenge 2010 (VOC2010) Results (2010)
11. Everingham, M., Van Gool, L., Williams, C., Winn, J., Zisserman, A.: The PASCAL Visual Object Classes Challenge 2011 (VOC2011) Results (2011)
12. Everingham, M., Van Gool, L., Williams, C., Winn, J., Zisserman, A.: The PASCAL Visual Object Classes Challenge 2011 (VOC2012) Results (2012)
13. Fei-Fei, L., Perona, P.: A Bayesian hierarchical model for learning natural scene categories. In: Proceedings of the IEEE Conference on Computer Vision and Pattern Recognition (CVPR) (2005)
14. Fei-Fei, L., Fergus, R., Torralba, A.: Recognizing and learning object categories. Short Course in the IEEE International Conference on Computer Vision (2009)
15. Felzenszwalb, P., Girshick, R., McAllester, D., Ramanan, D.: Object detection with discriminantly trained part-based models. IEEE Trans. Pattern Anal. Mach. Intell. **32**, 1627–1645 (2010)
16. Fergus, R., Perona, P., Zisserman, A.: Object class recognition by unsupervised scale-invariant learning. In: Proceedings of the IEEE International Conference on Computer Vision (ICCV) (2003)
17. Hillel, A.B., Weinshall, D.: Subordinate class recognition using relational object models. In: Proceedings of the Conference on Neural Information Processing Systems (NIPS) (2007)
18. Johnson, K.E., Eilers, A.T.: Effects of knowledge and development on subordinate level categorization. Cogn. Dev. **13**(4), 515–545 (1998)
19. Khosla, A., Xiao, J., Torralba, A., Oliva, A.: Memorability of image regions. In: Advances in Neural Information Processing Systems (NIPS), Lake Tahoe (2012)
20. Khosla, A., Yao, B., Jayadevaprakash, N., Fei-Fei, L.: Novel dataset for fine-grained image categorization. In: First Workshop on Fine-Grained Visual Categorization, IEEE Conference on Computer Vision and Pattern Recognition, Colorado Springs (2011)
21. Lazebnik, S.: Schmid, C., Ponce, J.: Beyond bags of features: spatial pyramid matching for recognizing natural scene categories. In: Proceedings of the IEEE Conference on Computer Vision and Pattern Recognition (CVPR) (2006)
22. Li, L.-J., Su, H., Xing, E., Fei-Fei, L.: Object bank: a high-level image representation for scene classification and semantic feature sparsification. In: Proceedings of the Conference on Neural Information Processing Systems (NIPS) (2010)
23. Lowe, David G.: Distinctive image features from scale-invariant keypoints. Int. J. Comput. Vision **60**(2), 91–110 (2004)
24. Moosmann, F., Triggs, B., Jurie, F.: Fast discriminative visual codebooks using randomized clustering forests. In: Proceedings of the Conference on Neural Information Processing Systems (NIPS) (2007)

25. Ojala, T., Pietikainen, M., Harwood, D.: Performance evaluation of texture measures with classification based on kullback discrimination of distributions. In: Proceedings of the IEEE International Conference on Pattern Recognition (ICPR) (1994)
26. Oliva, A., Torralba, A.: Modeling the shape of the scene: a holistic representation of the shape envelope. Int. J. Comput. Vision **42**(3), 145–175 (2001)
27. Shotton, J., Johnson, M., Cipolla, R.: Semantic texton forests for image categorization and segmentation. In: Proceedings of the IEEE Conference on Computer Vision and Pattern Recognition (CVPR) (2008)
28. Tu, Z.: Probabilistic boosting-tree: learning discriminative models for classification, recognition, and clustering. In: Proceedings of the IEEE International Conference on Computer Vision (ICCV) (2005)
29. van de Sande, K.E.A., Gevers, T., Snoek, C.G.M.: Evaluating color descriptors for object and scene recognition. IEEE Trans. Pattern Anal. Mach. Intell. **32**(9), 1582–1596 (2010)
30. van de Weijer, J., Schmid, C., Verbeek, J., Larlus, D.: Learning color names for real-world applications. IEEE Trans. Image Process. **18**(7), 1512–1523 (2009)
31. Wang, J., Yang, J., Yu, K., Lv, F., Huang, T., Gong, Y.: Locality-constrained linear coding for image classification. In: Proceedings of the IEEE Conference on Computer Vision and Pattern Recognition (CVPR) (2010)
32. Welinder, P., Branson, S., Mita, T., Wah, C., Schroff, F., Belongie, S., Perona, P.: Caltech-UCSD birds 200. Technical Report CNS-TR-201, Caltech (2010)
33. Yao, B., Fei-Fei, L.: Grouplet: a structured image representation for recognizing human and object interactions. In: Proceedings of the IEEE Conference on Computer Vision and Pattern Recognition (CVPR) (2010)
34. Yao, B., Fei-Fei, L.: Modeling mutual context of object and human pose in human-object interaction activities. In: Proceedings of the IEEE Conference on Computer Vision and Pattern Recognition (CVPR) (2010)
35. Yao, B., Jiang, X., Khosla, A., Lin, A.L., Guibas, L. J., Fei-Fei, L.: Human action recognition by learning bases of action attributes and parts. In: Proceedings of the IEEE International Conference on Computer Vision (ICCV) (2011)
36. Yao, B., Khosla, A., Fei-Fei, L.: Classifying actions and measuring action similarity by modeling the mutual context of objects and human poses. In: Proceedings of the International Conference on Machine Learning (ICML) (2011)
37. Yao, B., Khosla, A., Fei-Fei, L.: Combining randomization and discrimination for fine-grained image categorization. In: Proceedings of the IEEE Conference on Computer Vision and Pattern Recognition (CVPR) (2011)
38. Yao, B., Bradski, G., Fei-Fei, L.: A codebook-free and annotation-free approach for fine-grained image categorization. In: Proceedings of the IEEE Conference on Computer Vision and Pattern Recognition (CVPR) (2012)
39. Yao, B., Fei-Fei, L.: Action recognition with exemplar based 2.5D graph matching. In: Proceedings of the European Conference on Computer Vision (ECCV) (2012)

Part II
Human Attributes in Social Media Analytics

Chapter 6
Recognizing People in Social Context

Gang Wang, Andrew Gallagher, Jiebo Luo and David Forsyth

6.1 Introduction

This chapter deals with recognizing people in personal image collections by exploiting social relationship as an extra cue. Personal image collections now often contain thousands or tens of thousands of images. Images of people comprise a significant portion of these images. Consumers capture images of the important people in their lives in a variety of social situations. People who are important to the photographer often appear many times throughout the personal collection. Many factors influence the position and pose of each person in the image. We propose that familial social relationships between people, such as "mother-child" or "siblings", are one of the strong factors. For example, Fig. 6.1 shows two images of a family at two different events. We observe that the relative position of each family member is roughly the same. The position of a person relative to another is dependent on both the identity of the persons and the social relationship between them. To explore these ideas, we examine family image collections that have repeating occurrences of the same individuals and the social relationships that we consider are family relationships.

For family image collections, face recognition typically uses features based on facial appearance alone, sometimes including contextual features related to clothing [19, 24, 26]. In essence, that approach makes the implicit assumption that the identity

G. Wang (✉)
Electrical and Electronic Engineering, Nanyang Technological University, 50 Nanyang Avenue, Singapore 639798, Singapore
e-mail: wanggang@ntu.edu.sg

A. Gallagher
Cornell University, Ithaca, NY, USA

J. Luo
University of Rochester, Rochester, NY, USA

D. Forsyth
University of Illinois at Urbana-Champaign, Champaign, IL, USA

Y. Fu (ed.), *Human-Centered Social Media Analytics*,
DOI: 10.1007/978-3-319-05491-9_6, © Springer International Publishing Switzerland 2014

Fig. 6.1 Social relationships often exhibit certain visual patterns. For the two people in a wife–husband relationship, the face that is higher in the image is more likely to be the husband. The family members are in roughly the same position in the two images, even though the images are of two different events on different days. The inclination of people to be in specific locations relative to others in a social relationship is exploited in this work for recognizing individuals and social relationships

of a face is independent of the position of a face relative to others in the image. At its core, our work re-examines this assumption by showing that face recognition is improved by considering contextual features that describe one face relative to others in the image, and that these same features are also related to the familial social relationship.

In Wang et al. [27], we have developed a probabilistic model for representing the influence between pairwise social relationships, identity, appearance and social context. The experimental results show that adding social relationships results in better performance for face annotation. With the learned relationship models, we can in turn discover social relationships from new image collections where the social relationships are not manually annotated.

6.1.1 Survey of Related Work

Organizing consumer photo collections is a difficult problem. One effective solution is to annotate faces in photos and to search and browse images by people names [17]. Automatic face annotation in personal albums is a hot topic and attracts much attention [3, 31]. There has been pioneering work on using social cues for face recognition [11, 16, 25]. Stone et al. [25] works with strongly labeled data, and only has one type of relationship: friend or not. In comparison, we deal with weakly labeled images, and explicitly model a number of social relationships. In [11], the authors use the social attributes people display in pictures to better recognize genders, ages, and identities. However, [11] does not explicitly model different social relationships between people or recognize specific individuals. In [16], recognizing individuals improves by inferring facial attributes. We extend these works by using social relationships as attributes for pairs of people in an image for recognizing people and social relationships. In object recognition, people also explore attributes or auxiliary data for more accurate object detection and similarity measurement [28, 29].

Weak labeling is an area related to our work. In image annotation, ambiguous labels are related to generic object classes rather than names [1, 13]. Berg et al. [2] is an example where face recognition has been combined with weak labels. In that work, face models are learned from news pictures and captions about celebrities, but ordinary people and the social relationships between them are not considered.

Certainly, the use of social relationships for recognition constitutes a type of context. The social context is related to the social interactions and environment in which an image is captured, and consequently it is not necessarily inferred directly from image data. Our contextual features for describing the relative positions between pairs of people in an image are similar to the contextual features shown to be effective in general object recognition [6, 14, 20]. In these works, pairwise features enforce priors that, for example, make it unlikely for cows to appear in the sky. We show that our similar features are in fact also useful for improving person recognition and for identifying social relationships. In our work, social relationships act as a high-level context leveraged from human knowledge or human behavior. In this sense, it is similar to the context of [10, 21].

A number of papers on social relationship recognition and discovery were published after our original publication [27]. In [4], a new framework is developed to automatically discover informative social subgroups (e.g., family members and classmate). The results can be further used for service recommendation (e.g., recommending a family-friendly restaurant). Ding et al. [5] infer social relations among actors in a video by analyzing the interactions among the actors. Instead of recognizing general social relationships, another relevant line of work focuses on kinship relationship recognition, with examples including [7, 8, 22, 23, 30, 32].

6.2 Approach

The common method for providing labeled samples to construct a model of facial appearance for a specific individual involves asking a user to label a set of training faces for each person that is to be recognized. Then, a face model can be learned in a fairly straightforward manner. However, annotating specific faces in a manual fashion is a time-consuming task. In practice, tools such as Flickr, or Adobe Album are used by many consumers, but they only provide weak labels that indicate the presence of a person but not that person's location in the image. Appearance models can still be learned in this scenario, but the label ambiguity increases the learning difficulty. We assume this realistic weak-labeling scenario, similar to that of [2], and our model is used to disambiguate the labels, learn appearance models, and find the identity of persons in images that were not in the original training subset. Note also that other frameworks exist for minimizing the effort of the user by using active

Training input: Test:

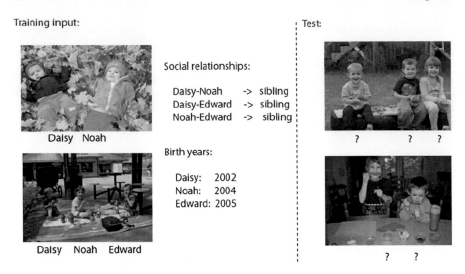

Fig. 6.2 In the training procedure, images are weakly labeled. Social relationships and birth years are annotated as input for learning social relationship models. In the recognition test procedure, the goal is to annotate faces present in images with names

learning to suggest samples to label [15, 26], and our model could be inserted into one of these frameworks.

The procedure is illustrated in Fig. 6.2. For each image, we only know there are N names annotated, which are written as $\{p_i, i = 1, \ldots, N\}$, but do not know the positions or scales of the corresponding faces. Most of faces are automatically detected from images, and we manually add missed faces since we are not studying face detection in this work. Each face is represented by Fisher subspace features [9]. Features of faces are written as $\{w_j, j = 1, \ldots, M\}$.

We train a face model for each individual. This requires establishing correspondences between names and faces in each training image. Social relationships are manually annotated by photo owners; the relationship between the ith and jth people is written as r_{ij}, a discrete variable over the nine pairwise social relationships that we consider. The labeling of this social relationship is reasonable and requires only a small amount of additional effort, because a given pairwise social relationship need be annotated only *once* for the entire personal collection. There are $N(N-1)/2$ possible pairwise relationships in one album with N people, but many pairs of people do not have direct relationships.

A specific social relationship usually exhibits common visual patterns in images. For example, in a "husband–wife" relationship, the husband is usually taller than the wife due to physical factors (e.g., the average adult male is 176.8 cm while the average female is 163.3 cm [18]). Of course, it is easy to find exceptions, and this is why our model relies not on "rules" that define the behavior of an individual or a person in a family relationship, but rather on probabilistic distributions of features f for particular social relationships.

Table 6.1 The notation for our model

p_i: the ith person name	P: all names
w_i: the feature representation of the ith face	W: all face features
t_i: the age of the ith person	T: all ages
r_{ij}: the social relationship between the ith and the jth person	R: all annotated relationships
f_{ij}: the social relationship features between the ith and the jth face	F: all social relationship features
A: the hidden variable which assigns names to faces	$A_i = j$: the ith name is assigned
θ: model parameters	to the jth face

Fig. 6.3 The graphical model. The notation is explained in Table 6.1

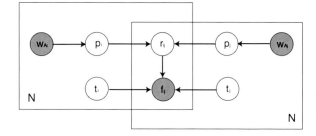

We extract features that reflect social relationships for each pair of faces i and j. The features describing the ith and jth face pair are written as f_{ij}. This feature vector represents the "social context" in our model. Note that even within a single social relationship, visual patterns are not time-invariant. For example, for "child–mother" relationship, when the child is an infant and the mother is in her 20s, the mother's face is physically larger than and generally positioned above the child's; but when the child grows, he or she may eventually have a larger face, be physically taller, and will no longer sit on the mother's lap. To accommodate the evolving roles within a social relationship, we allow the representation of social relationships for different age combinations. This requires that the collection owner provides approximate birth years for each person as illustrated in Fig. 6.2. In a training image, ages of people are written as $\{t_i, i = 1, \ldots, N\}$.

Given the above-defined notations, we then aim to maximize the conditional probability of labels given image observations $p(P, R, T \mid W, F)$, which can be rewritten as:

$$\frac{p(P, R, T, W, F)}{p(W, F)} \sim \sum_A p(P, R, T, W, F \mid A) p(A) \qquad (6.1)$$

A is a hidden variable that defines the correspondence between faces and names. $A_i = j$ denotes the ith name is assigned to the jth face. Given a specific A, the dependency between P, R, T, W and F is represented as shown in Fig. 6.3. We use a discriminative model to represent the appearance of each name (here we use a weighted KNN classifier due to its robustness, but note that a generative model such as a Gaussian mixture model is also applicable) and generative models for social relationships.

According to the graphical model, (6.1) can be written as:

$$\sum_{A} \prod_{i=1}^{N} p(p_i \mid w_{A_i}) \prod_{i=1,j=1}^{N} p(f_{A_i A_j} \mid r_{ij}, t_i, t_j) p(r_{ij} \mid p_i, p_j) p(A) \tag{6.2}$$

where w_{A_i} denotes the features of the face that is associated with the name p_i. r_{ij} is annotated for each pair of names p_i and p_j, so $p(r_{ij} \mid p_i, p_j)$ is 1 and neglected from now on. $p(p_i \mid w_{A_i})$ is calculated as:

$$p(p_i \mid w_{A_i}) = \frac{\sum_{l=1}^{L} p(p_i \mid w_l^{N_{A_i}})}{\sum_{i=1}^{N} \sum_{l=1}^{L} p(p_i \mid w_l^{N_{A_i}})} \tag{6.3}$$

where $w_l^{N_{A_i}}$ denotes the l nearest neighbor faces found for w_{A_i} in all the training images. $p(p_i \mid w_l^{N_{A_i}}) = 0$ if the image containing $w_l^{N_{A_i}}$ does not have the person p_i present. $\sum_i p(p_i \mid w_j) = 1$ is enforced in the training procedure.

$f_{A_i A_j}$ denotes the social relationship features extracted from the pair of faces A_i and A_j. We extract five types of features to represent social relationships, which are introduced in Sect. 6.3. The space of each feature is quantized to several discrete bins, so we can model $p(f_{A_i A_j}^k \mid r_{ij}, t_i, t_j)$ as a multinomial distribution, where k denotes the kth type of relationship features. For simplicity, these relationship features are assumed to be independent of each other, and $p(f_{A_i A_j} \mid r_{ij}, t_i, t_j)$ could simply be calculated as the product of the probability for each feature. However, we find that the features can be combined in smarter ways. By providing a learned exponent on each probability term, the relative importance of each feature can be adjusted. By learning the exponents with cross-validation on training examples, better performance is achieved.

There are many possible t_i and t_j pairwise age combinations, but we may only have a few training examples for each combination. However, visual features do not change much without a dramatic change of age. So we quantize each age t_i into five bins. The quantization partition points are $[0\ 2\ 17\ 35\ 60\ 100]$ years. Consequently, there are 25 possible pairwise age bin combinations. For each, we learn a multinomial distribution for each type of relationship feature. The multinomial distribution parameters are smoothed with a Dirichlet prior.

6.2.1 Learning the Model with EM

Learning is performed to find the parameters $\widehat{\theta}$:

$$\widehat{\theta} = \text{argmax}_\theta\, p(P, R, T \mid W, F; \theta) \tag{6.4}$$

θ contains the parameters to define $p(p \mid w)$ and $p(f \mid r, t)$. This cannot be learned with maximum likelihood estimation because of the hidden variable. Instead, we use the EM algorithm, which iterates between the E step and the M step. Initialization is critical to the EM algorithm. In our implementation, we initialize $p(p_i \mid w_j)$ with the parameters produced by the baseline model that omits the social relationship variables. The multinomial distribution is initialized as a uniform distribution.

In the E step, we calculate the probability of the assignment variable A given the current parameters θ^{old}. For a particular A^*, we calculate it as:

$$p(A^* \mid P, R, T, W, F; \theta^{\text{old}}) = \frac{p(P, R, T, W, F \mid A^*; \theta^{\text{old}})p(A^*; \theta^{\text{old}})}{\sum_A p(P, R, T, W, F \mid A; \theta^{\text{old}})p(A; \theta^{\text{old}})} \quad (6.5)$$

$p(P, R, T, W, F, A^*; \theta^{\text{old}})$ can be calculated according to (6.2). The prior distribution of A is simply treated as a uniform distribution. This needs to be enumerated over all the possible assignments. When there are a large number of people in images, it becomes intractable. We only assign one p_i to a w_j when $p(p_i \mid w_j)$ is bigger than a threshold, which is selected adaptively so that at most five p_i can be assigned to a single w_j. In this way, we can significantly reduce the number of possible A.

In the M step, we update the parameters by maximizing the expected likelihood function, which can be obtained by combing (6.2) and (6.5). There are two types of parameters, one to characterize $p(p \mid w)$ and the other one to characterize $p(f \mid r, t)$. In the M step, when updating one type of parameters using maximum likelihood estimation, the derivative doesn't contain the other type of parameters. Therefore, the updates of parameters for $p(p \mid w)$ and $p(f \mid r, t)$ are separate. When running the EM algorithm, the likelihood values do not change significantly after 5–10 iterations.

6.2.2 Inference

In the inference stage, we are given a test image containing a set of people (without any name label information), we extract their face appearance features W and relationship features F, then predict the names P. We use the relationship models to constrain the labeling procedure, so the classification of faces is not done based on facial appearance alone. This problem is equivalent to finding a one-to-one constraint A^* in the following way:

$$A^* = \text{argmax}_A\, p(A \mid P, R, W, F, T) \quad (6.6)$$

Here, P denotes all the names in the dataset. There would be too many possible A to evaluate and compare. We adopt a simple heuristic by only considering As which assign a name p to a face w when $p(p \mid w)$ is bigger than a threshold. This heuristic works well in our implementation.

6.3 Implementation Details

The appearance of each face is represented by projecting the original pixel values into a Fisher subspace learned from a held-out collection (containing no images in common with either the training set or the test set). Each face is represented as a Fisher discriminant space feature.

In our model, the social relationship variable r_{ij} is discrete over the space of pairwise social relationships. We represent the following nine familial social relationships between a pair of people:

mother-child	father-child	grandparent-child	husband-wife	siblings
child-mother	child-father	child-grandparent	wife-husband	

We consider relationships to be asymmetric (e.g., "mother-child" is different from "child-mother") because our objective is to identify the role of each person in the relationship. We use the following five types of observed appearance features to represent social relationships.

Height: the height difference is used as a feature. Very simply, we use the ratio of the difference y-coordinates of the two people's faces to the average face size of the faces in the image. The ratio is quantized to six bins.

Face size ratio: this feature is the ratio of the face sizes. We quantize the ratio to six bins.

Closeness: the distance of two people in an image can reveal something about their social relationship. We calculate the Euclidean distance between pair of people, normalized by the average face size. We quantize the distance to five bins.

We train gender and age classifiers based on standard methods, following the examples of [12, 16]. Two linear projections (one for age and one for gender) are learned and nearest neighbors (using Euclidean distance) to the query are found in the projection space.

Age difference: we use our age predictor to estimate the ages of people. This age difference, estimated purely from appearance, tells us some information about the social relationship. We quantize age into five ranges, so the age difference between two people has nine possibilities. The age difference relationship is modeled as a multinomial distribution over these nine bins.

Gender distribution: the appearance-based gender classifier helps to indicate the role of a person in a social relationship. For example, gender estimates are useful for distinguishing between a wife and husband (or more broadly a heterosexual couple). For each pair of people, there are four possible joint combinations of the genders.

Figure 6.4 demonstrates evidence of the dependence between social relationships and our features by showing the distribution of feature values given the social relationships, as learned from our training collections.

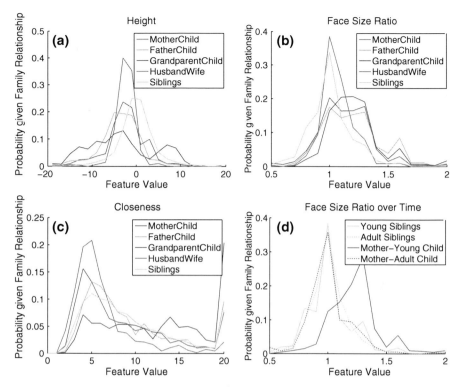

Fig. 6.4 Pairwise facial features are dependent on social relationships. From these plots, we see that parents' faces are usually above childrens' faces (**a**), that spouses' faces are usually about the same size, but are larger than children's (**b**), and spouses tend to be close together in an image (**c**). Note that we also model the changing nature of family relationships over time: a mother's face is larger than the child's when the child is young, but they are generally the same size when the child is an adult (**d**)

6.4 Experiments

We have conducted experiments that support our assertion that modeling social relationships provides improvements for recognizing people, and allows for the recognition of pairwise social relationships in new images.

In Sect. 6.4.1, we examine the task of identifying people through experiments on three personal image collections, each of which has more than 1,000 images and more than 30 distinct people. We show that significant improvement is made by modeling social relationships for face annotation on both datasets. We also investigate how different social relationships features help to boost the performance.

Furthermore, in Sect. 6.4.2, we show that learned social relationships models can be transferred across different datasets. Social relationships are learned on a personal image collection, and then social relationships are effectively classified in single images from unrelated separate image collections.

Portions of the data that we use in this section are publicly available at [27].

6.4.1 Recognizing People with Social Relationships

In the first experiment, a subset of images from a personal image collection is randomly selected as training examples, and weak name labels are provided for the identities of the people in the images. The remaining images comprise a test set for assessing the accuracy of recognizing individuals. Testing proceeds as follows: first, the correspondence between the names and the faces of the training images are found using the EM procedure from Sect. 6.2.1. Next, inference is performed (Sect. 6.2.2) to determine the most likely names assignment for each set of faces in each test image. The percentage of correctly annotated faces is used as the measure of performance. This measure is used to evaluate the recognition accuracy in the test set as well as in the training set.

The first collection has 1,125 images and contains 47 distinct people. These people have 2,769 face instances. The second collection contains 1,123 images, with 34 distinct people and 2,935 faces. The third collection has 1,117 images of 152 individuals and 3,282 faces. For each collection, we randomly select 600 images as training examples and the others as test examples. Each image contains at least two people. In total, these images contain 6,533 instances of 276 pairwise social relationships.

Improvement made by modeling social relationships: For comparison to our model that includes social relationships, we first perform experiments without modeling social relationships. In the training procedure, we maximize: $p(P \mid W) \sim \sum_A \prod_{i=1}^N p(p_i \mid w_{A_i}) p(A)$. Likewise, the EM algorithm is employed to learn model parameters.

Figure 6.5 shows that all datasets show improved recognition accuracy in both training and testing when social relationships are modeled. By modeling social relationships, better correspondence (i.e., disambiguation of the weak label names) in the training set is established. In collection 1, training set accuracy improves by 5.0 % by modeling social relationships, and test set identification improves by 8.6 % due to the improved face models as well as the social relationship models. Significant improvement is also observed in collection 2 in both the training (improves by 3.3 %) and test (improves by 5.8 %) sets. Collection 3 also shows improvement (by 9.5 % in training and by 1.8 % in testing) although the overall accuracy is lower, mainly because this collection contains many more unique people (152 people versus 47 and 34 in collections 1 and 2).

Figure 6.6 illustrates the improvement that modeling social relationships provides for specific test image examples. The faces in green squares are instances that are not correctly classified when the model ignores social relationships, but are corrected by modeling social relationships. We can see that these faces are surrounded by other people who have strong social relationships with, and the visual patterns between people are what is typically expected given their roles in the relationships. The faces in red squares are instances that are correctly classified when appearance alone is considered, but get confused by incorporating social relationships. This is because visual relationship patterns in these pictures are atypical of what is observed in most of other pictures. mother, so she is misclassified as her father, despite her childlike facial appearance.

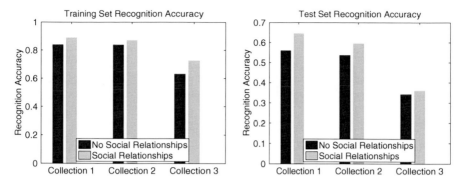

Fig. 6.5 Modeling social relationships improves recognition accuracy. The plots show the improvement in recognition accuracy for both the training set (*left*) and the test set (*right*) for two different image collections

Fig. 6.6 The faces in *green squares* are instances that are not correctly recognized without modelling social relationships, but are corrected by modeling social relationships. The faces in *red squares* are correctly recognized at first, but are misrecognized when social relationships are considered. The mistakes are sometimes due to an improbable arrangement of the people in the scene (e.g., the son on the father's shoulders in the *lower right*) that is not often observed in the training set. As another example, in the *middle* image of the second row, the daughter (closer to the camera) appears *taller* and has a *bigger* face size than her mother, so she is misclassified as her father, despite her childlike facial appearance. *Note* For color interpretation see online version

Effect of each social relationship feature: As described in Sect. 6.3, we use five features to encapsulate social relationships. We show how each type of relationship feature helps by in turn omitting all features except that one. The results are shown in Table 6.2. We observe that relative face size is the most helpful single feature, followed by age and gender. In general, including all features provides significant improvement over using any single feature and adding any single feature is better than using none at all. It is interesting to note that while our results concur with [16] in that we achieve improved face recognition by estimating age and gender.

Table 6.2 Person recognition accuracy in the test set improves for both collections by modeling social relationships using more features

	Without relationships	+Height	+Closeness	+Size	+Age	+Gender	+All
Collection 1	0.560	0.621	0.628	0.637	0.635	0.630	**0.646**
Collection 2	0.537	0.563	0.560	0.583	0.573	0.584	**0.595**
Collection 3	0.343	0.361	0.359	**0.362**	0.362	0.362	0.361
Overall mean	0.480	0.515	0.516	0.527	0.523	0.525	**0.534**

For example, "+height" means that only relative height feature is used, and the other features are omitted

6.4.2 Recognizing Social Relationships in Novel Image Collections

Our model explicitly reasons about the social relationships between pairs of people in images. As a result, the model has applications for image retrieval based on social relationships.

Social relationships are modeled with visual features such as relative face sizes and age difference, which are not dependent on the identities of people. This means social relationship models can be transferred to other image collections with different people. Consequently, the models learned from one image collection can be used to discover social relationships in a separate unrelated image collection with no labeled information at all. We perform two experiments to verify that we learn useful and general models for representing social relationships in images.

In the first experiment, we learn social relationship models from the training examples of collection 1, and classify relationships in collection 2. Because collection 2 contains no "grandparent–child" relationships, we limit the classified r_{ij} values to the other seven social relationships. The confusion matrix is shown in Fig. 6.7. Each row of this confusion matrix shows an actual class.

The averaged value of diagonals is 50.8 %, far better than random performance (14.3 %). We can see that the mistakes are reasonable. For example, "child-mother" is usually misclassified as "child-father" because the primary visual difference between "mother" and "father" is the gender, which may not be reliably detected from consumer images.

In a second experiment, we perform social relationship recognition experiments on the publicly released group image dataset [11]. First, we manually labeled relationships between pairs of people. A total of 708 social relationships were labeled, at most one relationship per image, and each of the three social relationships has over 200 samples. This dataset is used solely as a test set. The social relationship models are learned from collection 1 in the same weakly supervised learning fashion as before. The confusion matrix is shown in Fig. 6.7. The overall social relationship classification accuracy in this experiment is 52.7 %, again exceeding random classification 20.0 %. This performance is significant in that the entire model is trained

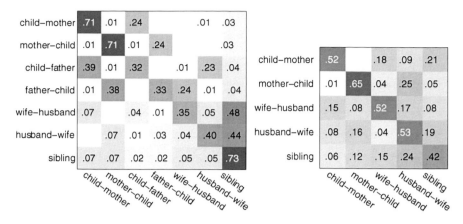

Fig. 6.7 The confusion matrix of social relationships classification. *Left* We learn social relationship models from collection 1 and test on the images of collection 2. *Right* We apply the learned social relationship models to a set of images from Flickr, and labeled as one of five social relationships. Both experiments show that social relationship models learned from one collection and transferable and useful for classifying social relationships in images containing strangers

Fig. 6.8 **a** social relationships classified as wife-husband. **b** social relationships classified as siblings. **c** social relationships classified as mother-child. Social relationship classification is accomplished from single images with our model, trained only with weak labels on a single, unrelated personal collection. Here, the task is to distinguish between the "wife–husband," "siblings," and "mother–child" relationships for each pair of *circled faces*. Incorrect classifications are outlined in *red*. *Note* For color interpretation see online version

on a single personal image collection with weak labels. Images classification results from the model are shown for three social relationships in Fig. 6.8.

6.5 Summary

This chapter discussed a model that incorporates pairwise social relationships such as husband-wife or mother-child for representing the relationship between people in a personal image collection. This model is motivated by the observation that the joint appearance between people in an image is associated with both their identities and the social relationship between the pair. We show experimentally several advantages of this representation. First, the model allows for establishing the correspondence between faces and names in weakly labeled images. Second, the identification of unknown faces in test images is significantly improved when social relationship inference is included. Third, social relationships models learned from the weakly labeled data are used to recognize social relationships in single previously unseen images.

We can foresee that our method can be applicable in many applications such as family album organization. Beyond, the way we develop to model the relationship between people can be potentially adapted to model the relationship between different object categories for other vision tasks such as scene understanding.

References

1. Barnard, K., Duygulu, P., Forsyth, D., de Freitas, N., Blei, D.M., Jordan, M.I.: Matching words and pictures. JMLR **3**, 1107–1135 (2003)
2. Berg, T., Berg, A., Edwards, J., Maire, M., White, R., Teh, Y.W., Learned-Miller, E., Forsyth, D.: Names and faces in the news. In Proceedings of the CVPR (2004)
3. Chen, L., Hu, B., Zhang, L., Li, M., Zhang, H.J.: Face annotation for family photo album management. IJIG **3**(1), 81–94 (2003)
4. Chen, Y.-Y., Hsu, W.H., Mark Liao H.-Y.: Discovering informative social subgraphs and predicting pairwise relationships from group photos. In: Proceedings of the 20th ACM International Conference on Multimedia, pp. 669–678. ACM (2012)
5. Ding, L., Yilmaz, A.: Inferring social relations from visual concepts. In IEEE International Conference on Computer Vision, pp. 699–706. IEEE (2011)
6. Divvala, S.K., Hoiem, D., Hays, J., Efros, A., Hebert, M.: An empirical study of context in object detection. In: Proceedings of the CVPR (2009)
7. Fang, R., Gallagher, A., Chen, T., Loui, A.: Kinship classification by modeling facial feature heredity. In: Proceedings of the ICIP (2013)
8. Fang, R., Tang, K.D., Snavely, N., Chen, T.: Towards computational models of kinship verification. In: 17th IEEE International Conference on Image Processing, pp. 1577–1580. IEEE (2010)
9. Gallagher, A., Chen, T.: Clothing cosegmentation for recognizing people. In: Proceedings of the CVPR (2008)
10. Gallagher, A., Chen, T.: Estimating age, gender, and identity using first name priors. In: Proceedings of the CVPR (2008)
11. Gallagher, A., Chen, T.: Understanding images of groups of people. In: Proceedings of the CVPR (2009)
12. Guo, G., Fu, Y., Dyer, C., Huang, T.: Image-based human age estimation by manifold learning and locally adjusted robust regression. In IEEE Trans. Image Process. **17**(7), 1111178–1111199 (2008)
13. Gupta, A., Davis, L.S.: Beyond nouns: Exploiting prepositions and comparative adjectives for learning visual classifiers. In: Proceedings of the ECCV (2008)
14. Hoiem, D., Efros, A.A., Hebert, M.: Putting objects in perspective. IJCV **80**(1), 3–15 (2008)
15. Kapoor, A., Hua, G., Akbarzadeh, A., Baker, S.: Which faces to tag: Adding prior constraints into active learning. In: Proceedings of the ICCV (2009)
16. Kumar, N., Berg, A., Belhumeur, P., Nayar, S.: Attribute and simile classifiers for face verification. In: Proceedings of the ICCV (2009)

17. Naaman, M., Yeh, R., Garcia-Molina, H., Paepcke, A.: Leveraging context to resolve identity in photo albums. In: Proceedings of the JCDL (2005)
18. National Center for Health Statistics. CDC growth charts, United States. http://www.cdc.gov/nchs/data/nhanes/growthcharts/zscore/statage.xls (2007)
19. O'Hare, N., Smeaton, A.: Context-aware person identification in personal photo collections. IEEE Trans MM (2009)
20. Parikh, D., Zitnick, L., Chen, T.: From appearance to context-based recognition: Dense labeling in small images. In: Proceedings of the CVPR (2008)
21. Pellegrini, S., Ess, A., Schindler, K., van Gool, L.: You'll never walk alone: Modeling social behavior for multi-target tracking. In: Proceedings of the ICCV (2009)
22. Shao, M., Xia, S., Fu. Kinship verification through transfer learning. In: IJCAI (2011)
23. Shao, M., Xia, S., Luo, J., Fu, Y.: Understanding kin relationships in a photo. In: IEEE Transactions on Multimedia (T-MM) (2012)
24. Song, Y., Leung, T.: Context-aided human recognition- clustering. In: Proceedings of the ECCV (2006)
25. Stone, Z., Zickler, T., Darrell, T.: Autotagging facebook: Social network context improves photo annotation. In: Proceedings of the CVPR Internet Vision Workshop (2008)
26. Tian, Y., Liu, W., Xian, R., Wen, F., Tang, X.: A face annotation framework with partial clustering and interactive labeling. In: Proceedings of the CVPR (2007)
27. Wang, G., Gallagher, A., Luo, J., Forsyth, D.: Seeing people in social context: Recognizing people and social relationship. In: Proceedings of the ECCV, pp. 169–182. Springer (2010)
28. Wang, G., Forsyth, D.: Joint learning of visual attributes, object classes and visual saliency. In: IEEE 12th International Conference on Computer Vision, 2009, pp. 537–544. IEEE (2009)
29. Wang, G., Hoiem, D., Forsyth, D.: Learning image similarity from flickr groups using fast kernel machines. IEEE Trans. Pattern Anal. Mach. Intell. 34(11), 2177–2188 (2012)
30. Xia, Siyu, Shao, Ming, Luo, Jiebo: Understanding kin relationships in a photo. IEEE Trans. Multimedia 14(4), 1046–1056 (2012)
31. Zhao, M., Teo, Y.W., Liu, S., Chua, T., Jain, R.: Automatic person annotation of family photo album. Lect. Notes Comput. Sci. 4071, 163 (2006)
32. Zhou, X., Hu, J., Lu, J., Shang, Y., Guan, Y.: Kinship verification from facial images under uncontrolled conditions. In: Proceedings of the 19th ACM International Conference on Multimedia, pp. 953–956. ACM (2011)

Chapter 7
Female Facial Beauty Attribute Recognition and Editing

Jinjun Wang, Yihong Gong and Douglas Gray

7.1 Introduction

Recently, automatic human attributes recognition has caught a lot of researchers attentions due to its huge application potentials, such as human–computer interaction, surveillance, content based search, targeted advertising, etc [1]. These attributes include not only the popular face detection [2] and face recognition [3], but also other attributes such as age [4–9], gender [10–15], attractiveness score [16–19], ethnicity [20–22], emotional state, etc. Although non-intrusive recognition methods are still very challenging due to the large amount of ethnic and race variations, recent advances in computer vision technologies have enabled some very sophisticated and advanced algorithms to utilize the rich image information from cameras to solve this problem. In this chapter, we are most interested in performing the task using facial images [23] due to the ease of collecting facial images and also the strong representation power by facial images.

It is well known that human faces convey lots of high-level semantic information about the identification of human attributes. The major step for recognizing these attributes is to obtain an "Attribute Classifier" which classifies the human attributes from the facial images. This chapter is devoted to this interesting and complicated problem. We present a system [24] that can first detect faces from natural scene and then perform recognition of the beauty score, based on a modified Convolutional Neural Network [25]. We believe that the work enriched the experiences of AI research toward building generic intelligent systems.

J. Wang (✉) · Y. Gong
Xi'an Jiaotong University, 28 West Xianning Road, Xi'an 710049, Shaanxi, China
e-mail: jinjun@mail.xjtu.edu.cn

D. Gray
A9.com, 130 Lytton Ave, Palo Alto, CA 94301, USA

Y. Fu (ed.), *Human-Centered Social Media Analytics*,
DOI: 10.1007/978-3-319-05491-9_7, © Springer International Publishing Switzerland 2014

7.2 Related Work

In order to interact socially, there is vast amount of literature on cognitive psychology attesting capabilities of human at identifying faces. Many of the works to date have been on identifying age, gender, attractiveness score, ethnicity, etc. This section lists some representative works in each respective field.

7.2.1 Age Recognition

Age classification of person from the digital image is challenging, as highlighted by numerous authors in their research. Although accuracy like human beings is not yet achievable, existing techniques can give reasonable performance on small dataset. To give some examples, Asuman et al. [5] used concatenated LBP feature with k-nearest neighbor classifier and achieved 80% age recognition accuracy on FERET and FG-NET Database. Chen et al. [4] extracted features like gray level image and edge image with support vector machine (SVM) to estimate the level of age group and achieved 87% accuracy. Wang et al. [6] considered age estimation as a multi-class classification problem, where aging features, including Gabor and LBP were fused under the Adaboost framework to boost ECOC (Error correcting output coding) codes. The ECOC code is then input to an SVM classifier to recognize age groups. To reduce memory consumption, PCA was applied on LBP and Gabor features respectively before combing into single vector. Their system achieved 85% accuracy on FERET and FG-NET Database. Readers can refer to [26] for a more complete survey in this domain.

There are also a few research works focusing on studying the relationship between multiple attributes. For instances, Guo et al. [9] evaluated gender categorization with age variation. LBP and HOG features were used for gender classification in age variations, and accuracy of 86.55% (Young), 95.03% (Adult) and 89.04% (senior) were reported. Iga et al. [8] applied graph matching method to detect the position of the face, and then used geometric arrangement of hair and mustache for gender estimation, and used texture spots, wrinkles, and flabs for age estimation. Young et al. [7] combined a primary type of features such as eyes, nose, mouth, chin, virtual top of the head and sides of the face, and a secondary type of features, such as wrinkle, to determine the face into one of the four classes, babies, young, adults and seniors.

7.2.2 Gender Recognition

Gender recognition is another challenging task in computer vision, although it is very easy for human. Reported works in this field can be broadly divided into two major groups: Feature extraction and transformation method, and classification algorithm.

For feature and feature selection, Tran et al. [10] introduced a novel method for gender recognition by using 2D principal component analysis based feature extraction with SVM classifier. This approach achieved 4.51 % gender recognition error rate on FERET dataset. Lu et al. [11] applied ICA for Harr features and Gabor features from ellipse face images for feature selection, and used SVM as classifier. The method achieved better gender recognition accuracy. Brunelli et al. [12] used a set of 16 geometric features per image to train two competing RBF networks, one for male and other for female, and the classification rate was 79 % on 168 images. Lian and Lu [13] presented a multi-view gender classification system that considered both shape and texture information to represent facial images. The facial area was divided into small regions, from which LBP histograms are extracted and concatenated to efficiently represent the face. Sun et al. [14] proposed a genetic feature subset selection algorithm where PCA was firstly used to convert each image into a low-dimensional feature representation, and then genetic algorithms was applied to further select a subset of features from the low-dimensional space by disregarding certain eigenvectors that do not seem to encode important gender information. Naive Bayes, Neural Network, SVM, and LDA classifiers were then compared. Ku et al. [15] extracted a local structure features computed from a 3×3 pixel neighborhood using a modified version of the census transform with four stage classifier.

There are also several works focused on classification model and algorithm for gender recognition. For instances, Gutta et al. [27] presented a system with an ensemble of RBF networks and inductive decision trees. Experiments on FERET Database reported 96 % accuracy for gender classification. Lin and Wang [28] presented the fuzzy support vector machine with good generalization ability, where fuzzy membership functions were assigned to each input face feature, and such fuzzification demonstrated clear contributions at learning the decision surface. Moghaddam and Yang [29] used SVM with radial basis function kernels to classify low resolution (12×12) "Thumbnail" faces, where 1755 faces from the FERET dataset were evaluated, and the gender classification accuracy was 96 %.

7.2.3 Beauty Score Recognition

The notion of beauty has been an ill defined abstract concept for many years. Serious discussion of beauty has traditionally been the purview of artists and philosophers. It was not until the latter half of the twentieth century that the concept of facial beauty was explored by social scientists [16] and not until very recently that it was studied by computer scientists [17]. The social science approach to this problem can be characterized by the search for easily measurable and semantically meaningful features that are correlated with a human perception of beauty. Alley and Cunningham showed that averaging many aligned face images together produced an attractive face, but that many attractive faces were not at all average [18]. Grammer and Thornhill showed that facial symmetry can be related to facial attractiveness [19]. Since that time, the need for more complex feature representations has shifted research in this area to computer scientists.

Most computer science approaches to this problem can be described as geometric or landmark feature methods. A landmark feature is a *manually* selected point on a human face that usually has some semantic meaning such as *right corner of mouth* or *center of left eye*. The distances between these points and the ratios between these distances are then extracted and used for classification using some machine learning algorithms. While there are some methods of extracting this information automatically [30] most previous work relies on a very accurate set of dense manual labels, which are not currently available. Furthermore most previous methods are evaluated on relatively small datasets with different evaluation and ground truth methodologies. In 2001 Aarabi et al. built a classification system based on eight landmark ratios and evaluated the method on a dataset of 80 images rated on a scale of 1–4 [17]. Eisenthal et al. assembled an ensemble of features that included landmark distances and ratios, an indicator of facial symmetry, skin smoothness, hair color, and the coefficients of an eigenface decomposition [31]. Their method was evaluated on two datasets of 92 images each with ratings 1–7. Kagian et al. later improved upon their method using an improved feature selection method [32]. Most recently Guo and Sim have explored the related problem of automatic makeup application [9], which uses an example to transfer a style of makeup to a new face.

7.2.4 Ethnicity Recognition

Ethnicity is an important demographic attribute, and automatic classification of ethnicity has presented yet another challenge in face processing. Several authors have attempted this research problem. To list some examples, Hosoi et al. [22] proposed to use Gabor wavelets transformation with retina sampling to extract key facial features, and then used SVM for ethnicity classification. Lin et al. [21] presented the MM-LBP (Multi-scale Multi-ratio LBP) method where LBP histograms were extracted from multi-scale, multi-ratio rectangular regions over both texture and range images, and then Adaboost was utilized to construct a strong classifier for ethnicity recognition. Ou et al. [20] presented a real-time race classification system that achieved 82.5 % ethnicity classification accuracy in 750 face images from the FERET dataset.

7.3 Facial Beauty Attribute Recognition

Most of the above-mentioned systems are usually restricted to a very small and meticulously prepared subset of the population, such as uniform ethnicity, age, expression, pose and/or lighting conditions, etc. The images are studio-quality photos taken by professional photographers. As another limitation, all these methods are not fully automatic recognition systems, because they rely heavily on the accurate manual localization of landmark features and often ignore the image itself once they are collected.

We have attempted to solve the problem with fewer restrictions on the data and a ground truth rating methodology that produces an accurate ranking of the images in the data set. We have collected images of frontal female faces with few restrictions on ethnicity, lighting, pose, or expression. Most of the face images are cropped from low-quality photos taken by cell phone cameras. Some examples can be found in Fig. 7.4. Because of the heavy cost of labeling landmark features on such a large data set, in this chapter, we solely focused on methodologies which do not require these features. Furthermore, although landmark features and ratios appear to be correlated with facial attractiveness, it is yet unclear to what extent human brains really use these features to form their notion of facial beauty. In this chapter, we test the hypothesis if a biologically inspired learning architecture can achieve a near human-level performance on this particular task using a large data set with few restrictions.

In this section, we present a method [24] of quantifying and predicting female facial beauty score using an automatically learned appearance model, as compared to a manual geometric model. To the best of our knowledge, it is the first work to test if a Hubel-Wiesel model can achieve a near human-level performance on the task of scoring female facial attractiveness.

7.3.1 Preprocessing of Pairwise Labeling

There are several kinds of ratings that can be collected for this task as labeling. The most popular one is the absolute rating where a user is presented with a single image and asked to give a score, typically between 1 and 10. Most previous work has used some version of absolute ratings usually presented in the form of a Likert scale [33], where the user is asked about the level of agreement with a statement. This form of rating requires many users to rate each image such that a distribution of ratings can be gathered and averaged to estimate the true score. This method is not ideal because each user will have a different system of rating images and a user's rating of one image may be affected by the rating given to the previous image, among other things.

Another method [34] was to ask a user to sort a collection of images according to some criteria. This method is likely to give reliable ratings but it is challenging for users to sort a large dataset as it requires considering all the data at once.

The more practical method is to present a user with pair of images and ask which is more attractive. This method presents a user with a binary decision, which we have found can be made more quickly than an absolute rating. This is the method that we have chosen to label our data. Although pairwise ratings are easy to collect, in order to use them for building a scoring system we need to convert the ratings into an absolute score for each image. To convert the scores from pairwise to absolute, we minimize a cost function defined such that as many of the pairwise preferences as possible are enforced and the scores lie within a specified range. Let $s = \{s_1, s_2, \ldots, s_N\}$ be the set of all scores assigned to images 1 to N. We formulate the problem into minimizing the cost function:

$$J(s) = \sum_{i=1}^{M} \phi(s_i^+ - s_i^-) + \lambda s^T s, \qquad (7.1)$$

where (s_i^+/s_i^-) represents the ith comparison that consists of two scores s_i^+ and s_i^- for the two respective images. We require s_i^+ to be greater than s_i^-, and have accordingly defined the cost function $\phi(d) = e^{-d}$ to penalizes images that have scores which disagree with one of M pairwise preferences. In fact, this function can be any monotonically increasing cost function such as the hinge loss, which may be advisable in the presence of greater labeling noise. λ is a regularization constant that controls the range of final scores. A gradient descent approach is then used to minimize this cost function. This iterative approach was chosen because when we receive new labels, we can quickly update the scores without resolving the entire problem. Our implementation is built on a web server which updates the scores in real-time as new labels are entered.

7.3.2 Active Learning

When our system is initialized, all images have a zero score and image pairs are presented to users at random. However, as many comparisons are made and the scores begin to disperse, the efficacy of this strategy decays. The reason for this is due in part to labeling noise. If two images with very different scores are compared, it is likely that the image with the higher score will be selected. If this is the case, since the difference between the images are large, we learn almost nothing from this comparison. However, if the user accidentally clicks on the wrong image, this can have a very disruptive effect on the accuracy of the ranking.

For this reason, we use a relevance feedback approach to select image pairs to present to the user. We first select an image at random with probability inversely proportional to the number of ratings r_i, it has received so far.

$$p(I_i) = \frac{(r_i + \epsilon)^{-1}}{\sum_{j=1}^{N}(r_j + \epsilon)^{-1}}. \qquad (7.2)$$

We then select the next image with probability that decays with the distance to first image score.

$$p(I_i|s_1) = \frac{\exp(-(s_1 - s_i)^2/\sigma^2)}{\sum_{j=1}^{N} \exp(-(s_1 - s_j)^2/\sigma^2)}, \qquad (7.3)$$

Fig. 7.1 Simulation results for converting pairwise preferences to an absolute score

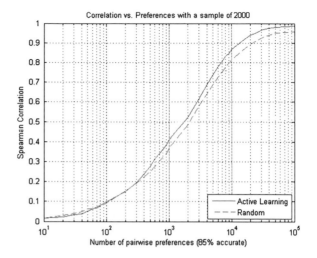

where σ^2 is the current variance of s. This approach is similar to the tournament sort algorithm and has significantly reduced the number of pairwise preferences needed to achieve a desired correlation of 0.9 (15 vs. 20k). Figure 7.1 shows the results of a simulation similar in size to our dataset. In this simulation 15 % of the preferences were marked incorrectly to reflect the inherent noise in collecting preference data.

7.3.3 Beauty Score Recognition

Given a set of images and associated beauty scores, our task is to train a regression model that can predict those scores. We adopt a predictive function that models the relationship between an input image I and the output score s, and learn the model in the following way

$$\min_{\mathbf{w},\theta} \sum_{i=1}^{N} (s_i - y_i)^2 + \lambda \mathbf{w}^T \mathbf{w},$$

$$\text{s.t.} \quad y_i = \mathbf{w}^T \Phi(I_i; \theta) + b, \tag{7.4}$$

where I_i is the raw-pixel of the ith image represented by size 128×128 in YCbCr colorspace, $\mathbf{w}$ is a D-dimensional weight vector, b is a scalar bias term, λ is a positive scalar fixed to be 0.01 in our experiments. As a main difference from the previous work, here we use $\Phi(\cdot)$ to directly operate on raw pixels I for extracting visual features, and its parameters θ are *automatically learned from data* with no manual efforts. In our study, we investigated the following special cases of the model, whose differences are the definition of $\Phi(I; \theta)$:

- **Eigenface Approach**: The method has been used for facial beauty prediction by [31], perhaps the only attempt so far requiring no manual landmark features. The method is as follows. We first run singular value decomposition (SVD) on the input training data $[I_1, \ldots, I_N]$ to obtain its rank D decomposition $\mathbf{U}\Sigma\mathbf{V}^T$, and then set $\theta = \mathbf{U}$ as a set of linear filters to operate on images so that $\Phi(I_i; \theta) = \mathbf{U}^T I_i$. We tried various D among $\{10, 20, 50, 100, 200\}$ and found that $D = 100$ gave the best performance.
- **Single Layer Model**: In contrast to Eigenface that uses *global* filters of receptive field 128×128, this model consists of 48 *local* linear filters of 9×9 size, each followed by a nonlinear logistic transformation. The filters convolute over the whole image and produce 48 feature maps, which were then down sampled by running max operator within each nonoverlapping 8×8 region, and thus reduced to 48 smaller 15×15 feature maps. The results serve as the outputs of $\Phi(I_i; \theta)$.
- **Two Layer Model**: We further enrich the complexity of $\Phi(I_i; \theta)$ by adding one more layer of feature extraction. In more details, in the first layer the model employs 16 filters of 9×9 size on the luminance channel, and 8 filters of 5×5 size on a downsampled chrominance channel; in the second layer, 24 filters of 5×5 size are connected to the output of the previous layer, followed by max downsampling by a factor of four.
- **Multiscale Model**: The model is similar to the single layer model, but with three additional convolution/downsampling layers. A diagram of this model can be found in Fig. 7.2. This model has 2974 tunable parameters, an order of magnitude less than a typical model trained for the task of face verification in [35].

In each of our models, every element of each filter is a learnable parameter, e.g. if our first layer has eight 5×5 filters, then there will be 200 tunable parameters in that layer. As we can see, these models represent a family of architectures with gradually increased complexities: *from linear to nonlinear, from single layer to multilayer, from global to local, and from course to fine* feature extractions. In particular, the employed max operator makes the architecture more local and partially scale-invariant, which is particularly useful in our case to handle the diversity of natural facial photos. The architectures can all be seen as a form of convolutional neural network [35, 36] that realizes the well-known Hubel-Wiesel model [37] inspired by the structure and functionalities of the visual cortex.

These systems were trained using stochastic gradient descent with a quadratic loss function. Optimal performance on the test set was usually found within a few hundred iterations, models with fewer parameters tend to converge faster both in iterations and computation time. We have tested many models with varying detailed configurations, and found in general that the number and size of filters are not crucial but the number of layers are more important—$\Phi(I_i; \theta)$ containing four layers of feature extraction generally outperformed the counterparts with fewer layers.

Fig. 7.2 An overview of the organization of our **multiscale** model. The first convolution is only performed on the luminance channel. Downsampled versions of the original image are fed back into the model at lower levels. *Arrows* represent downsampling, *lines* represent convolution and the *boxes* represent downsampling with the max operator. Feature dimensions are listed on the *left* (height × width × channels)

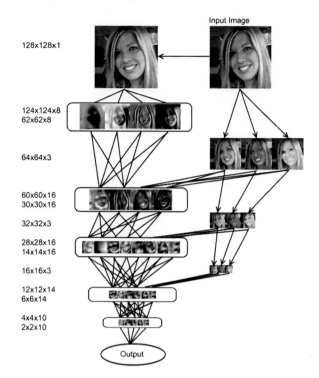

7.3.4 Beautify and Beastify the Original Image

In training a neural network, the back-propagation algorithm has been widely adopted by propagating the gradient of the final error function back through each layer in a network so that the gradient of each weight can be calculated w.r.t. the final error function. When a neural network is trained, the training input and associated labels are fixed, and the weights are iteratively optimized to reduce the error between the prediction and the true label.

A regularized cost function w.r.t. a desired score $(s^{(d)})$ and the corresponding gradient descent update can be written as:

$$J(I_t) = \phi(s_t - s^{(d)}) + \lambda\phi(I_t - I_0), \qquad (7.5)$$

and

$$I_{t+1} = I_t - \omega\left(\frac{\partial I_t}{\partial s} + \lambda(I_t - I_0)\right). \qquad (7.6)$$

where I_0 denotes the initial image, and I_t is the image obtained after iteration t. In our implementation, we use $\phi(x) = x^2$ and use different values of λ for the luminance and chrominance color channels. ω controls the image updating speed.

The gradient descent approach can be used both to *beautify* and *beastify* the original image. If we vary the regularization parameters and change the sign of the derivative, we can visualize the image manifold induced by the optimization. Figure 7.6 shows how specific features are modified as the regularization is relaxed.

We propose the *dual problem*. Given a trained neural network, fix the weights, set the gradient of the prediction to a fixed value and back propagate the gradient all the way through the network to the input image. This gives the derivative of the image w.r.t. the concept the network was trained with. This information is useful for several reasons. Most importantly, it indicates the regions of the original image that are most relevant to the task at hand. Additionally, the sign of the gradient indicates whether increasing the value of a particular pixel will increase or decrease the network output, meaning we can perform a gradient descent optimization on the original image.

7.4 Experimental Results

In order to make a credible attack on this problem we require a large dataset of high quality images, each labeled with a beauty score. For the purpose we resort to the popular website HOTorNOT[1] that has millions of images and billions of ratings. Users who submit their photo to this site waive their privacy expectations and agree to have their likeness criticized. Since the ratings associated with images in this dataset were collected from images of people as opposed to faces, we ran face detection software on a subset of images from this website and produced a dataset of 2,056 images and collected ratings of our own from 30 labelers.

7.4.1 Prediction Results

A full and complete comparison with previous work would be challenging both to perform and interpret. Most of the previous methods that have been successful relied on many manually marked landmark features, the distances between them, the ratios between those distances, and other hand crafted features. Manually labeling every image in our dataset by hand would be very costly so we will only compare with methods which do not require landmark features. As of the time of publication, the only such method is the **eigenface** approach used in [31].

We compare the four learning methods described in Sect. 7.3.3 based on the 2056 female face images and the absolute scores computed from pairwise comparisons. For each method, we investigate its performance on faces with and without face alignment. We perform alignment using the unsupervised method proposed in [38]. This approach is advantageous because it requires no manual annotation. In all the experiments, we fixed the training set to be 1028 randomly chosen images and used

---

[1] http://www.hotornot.com/

Fig. 7.3 A Scatter plot showing actual and predicted scores with the corresponding faces

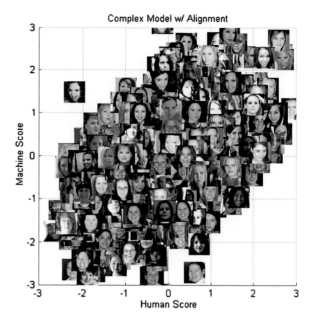

Complex Model w/ Alignment

Table 7.1 Correlation score of different methods

Method	Correlation w/o alignment	Correlation w/ alignment
Eigenface	0.134	0.180
Single layer model	0.403	0.417
Two layer model	0.405	0.438
Multiscale model	0.425	0.458

the remaining 1028 images for test. We show in Fig. 7.3 a scatter plot of some facial beauty score prediction example results.

Next, the pearson's correlation coefficient is used to evaluate the alignment between the machine-generated score and the human absolute score on the test data. Table 7.1 shows a comparison between the four methods—**eigenface, single layer, two layer** and **multiscale** models. We can see a significant improvement in the performance with alignment for the **eigenface** approach and a slight improvement for the hierarchal models. This discrepancy is likely due to the translation invariance that is introduced by the local filtering and downsampling with the max operator over multiple levels, as was first observed by [36]. Another observation is, with more layers being used, the performance improves. We note that **eigenface** produced a correlation score 0.40 in [31] on 92 studio quality photos of females with similar ages and the same ethnicity origins, but resulted very poor accuracy in our experiments. This shows that the large variability of our data significantly increased the difficulty of appearance-based approaches.

Fig. 7.4 The *top* (**a/b**) and *bottom* (**c/d**) *eight* images from our dataset according to human ratings (**a/c**) and machine predictions (**b/d**)

Though the Pearson's correlation provides a quantitative evaluation on how close the machine-generated scores are to the human scores, it lacks of intuitive sense about this closeness. In Fig. 7.3 we show a scatter plot of the actual and predicted scores for the **multiscale** model on the aligned test images. This plot shows both the correlation found with our method and the variability in our data. One way to look at the results is that, if without knowing the labels of axes, it is quite difficult to tell which dimension is by human and which by machine. We highly suggest readers to try such a test on Fig. 7.3 with an enlarged display.

Figure 7.4 shows the top and bottom eight images according to humans and the machine. Note that the ground truth for our training was generated with around 10^4 pairwise preferences, which is not sufficient to rank the data with complete accuracy. However, the notion of complete accuracy is something that can only be achieved for a single user, as no two people have the same exact preferences.

7.4.2 Discussion

With so much variability it is difficult to determine what features are being used for prediction. In this section, we discuss a method of identifying these features to better understand the learned models. One of the classic criticisms of the hierarchical model and neural networks in general, is the *black box* problem. That is, what features are we using and why are they relevant? This is typically addressed by presenting the convolution filters and noting their similarity to edge detectors (e.g., gabor filters). This was interesting when it was presented the first time, but by now everyone in the

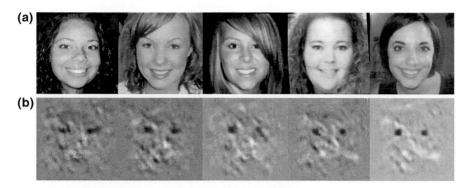

Fig. 7.5 Several faces (**a**) with their beauty derivative (**b**). These images are averaged over 10 gradient descent iterations and scaled in the colorspace for visibility

Fig. 7.6 The manifold of beauty for two images. **a** From *left* (beast) to *right* (beauty) we can see how the combination of (λ_Y/λ_C) for the luminance channel (λ_Y) and the chrominance channel (λ_C) controls the amount of modification. Specific features from **a** are Eyes (**b**) and Noses (**c**)

community knows that edges are important for almost every vision task. We attempt to address this issue using a logical extension to the back-propagation algorithm.

As explained in Sect. 7.3.4, we have proposed to use the *derivative of beauty* to beautify or beastify the original face. Figure 7.5 shows several example images and their respective gradients with respect to beauty for the **multiscale** model trained on aligned images. This in fact can also be a good indicator of what features are important for the prediction of beauty score.

From Fig. 7.5, the first observation is that women often wear dark eye makeup to accentuate their eyes. This makeup often has a dark blue or purple tint. We can see this reflected on the extremes of Fig. 7.6c. In Fig. 7.6b, the eyes on the bottom are dark blue/purple tint while the eyes on the top are bright with a yellow/green tint.

The second observation is that large noses are generally not very attractive. If we again look at the extremes of Fig. 7.6c we can see that the edges around the nose on the right side have been smoothed, while the same edges on the left side have been accentuated.

The final observation is that a bright smile is attractive. Unfortunately, the large amount of variation in facial expressions and mouth position in our training data

Fig. 7.7 The average face image (**a**), beautified images (**b**) and beastified images (**c**). The *x* axis represents changes in the luminance channel, while the *y* axis represents changes in the chrominance channels

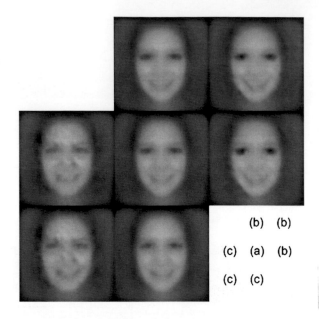

(b) (b)

(c) (a) (b)

(c) (c)

leads to artifacts in these regions such as in the extremes of Fig. 7.6. However, when we apply these modifications to the average image in Fig. 7.7, we can see a change in the perceived expression.

One further discussion is that, in the study of facial beauty, averaged faces are known to be attractive [18]. This is known as the averageness hypothesis. The average face from the dataset, presented in Fig. 7.7, has a score of 0.026. The scores returned by the proposed model are all zero mean, indicating that the average face is only of average attractiveness. This would seem to contradict the averageness hypothesis, however since the dataset presented here was collected from a pool of user submitted photos, it does not represent a truly random sampling of female faces (i.e., it may have a positive bias).

Previously, averageness, symmetry, and face geometry are the only definable features that have been shown to be correlated with facial attractiveness. This chapter presents evidence that many of the cosmetic products used by women to darken their eyes and hide lines and wrinkles are in fact attractive features.

7.5 Summary

We have presented a method of both quantifying and predicting female facial beauty score attribute using a hierarchical feed-forward model. Our method does not require landmark features, which makes it complimentary to the traditional geometric approach [17, 31, 32, 39, 40] when the problem of accurately estimating landmark feature locations is solved. The system has been evaluated on a more realistic dataset that is an order of magnitude larger than any previously published results. It has been shown that in addition to achieving a statistically significant level of correlation with human ratings, the features extracted have semantic meaning. We believe that the work enriches the experience of AI research toward building generic intelligent systems.

References

1. Shewaye, T.N.: Age group and gender recognition from human facial images. In: IEEE CVPR (2012)
2. Yang, M.-H., Kriegman, D., Ahuja, N.: Detecting faces in images: a survey. In: IEEE Transactions on Pattern Analysis and Machine Intelligence, pp. 34–58 (2002)
3. Zhang, C., Zhang, Z.: Survey of recent advances in face detection. Microsoft Technical Report MSR-TR-2010-66 (2010)
4. Chen, Y., Han, M., Song, K., Ho, Y.: Image-based age-group classification design using facial features. In: System Science and Engineering (ICSSE), pp. 548–522 (2010)
5. Asuman, G., Nabiyev, V.: Automatic age classification with LBP. In: Computer and Information Sciences (ISCIS), pp. 1–4 (2008)
6. Wang, J., Yau, W., Wang, H.: Age categorization via ECOC with fused gabor and LBP features. In: Workshop Applications of Computer Vision (WACV), pp. 1–6 (2009)
7. Kwon, Y., Lobo, N.: Age classification from facial images. J. Comput. Vis. Image Underst. **74**(1), 1–21 (1999)
8. Iga, R., Izumi, K., Hayashi, H., Fukano, G., Ohtani, T.: A gender and age estimation system from face images. In: Proceeding of SICE 2003 Annual Conference, pp. 332–339 (2003)
9. Guo, D., Sim, T.: Digital face makeup by example. In: IEEE Computer Society Conference on Computer Vision and Pattern Recognition (2009)
10. Bui, L., Tran, D., Huang, X., Chetty, G.: Face gender recognition based on 2D principal component analysis and support vector machine. In: 4th International Conference on Network and System Security (NSS), pp. 579–582 (2010)
11. Lu, H., Lin, H.: Gender recognition using adaboosted feature. In: Third International Conference on Natural Computation, vol. 2, pp. 646–650 (2007)
12. Brunelli, R., Poggio, T.: HyperBF networks for gender classification. In: Proceedings of the DARPA Image Understanding Workshop, pp. 311–314 (1992)
13. Lian, H., Lu, B.: Multi-view gender classification using local binary patterns and support vector machines. In: Advances in Neural Networks (2006)
14. Sun, Z., Bebis, G., Yuan, X., Louis, S.: Genetic feature extraction for gender classification: a comparison study. In: IEEE Workshop on Applications of Computer Vision (2002)
15. Ku, C., Ernst, A.: Face detection and tracking in video sequences using the modified census transformation. Image Vis. Comput. **24**, 564–572 (2006)
16. Cross, J., Cross, J.: Age, sex, race, and the perception of facial beauty. Dev. Psychol. **5**, 433–439 (1971)
17. Aarabi, P., Hughes, D., Mohajer, K., Emami, M.: The automatic measurement of facial beauty. In: IEEE International Conference on Systems, Man, and Cybernetics, vol. 4 (2001)
18. Alley, T., Cunningham, M.: Averaged faces are attractive, but very attractive faces are not average. Psychol. Sci. **2**, 123–125 (1991)
19. Grammer, K., Thornhill, R.: Human (*Homo sapiens*) facial attractiveness and sexual selection: the role of symmetry and averageness. J. Comp. Psychol. **108**, 233–242 (1994)
20. Ou, Y., Wu, X., Qian, H., Xu, Y.: A real time race classification system. In: IEEE International Conference on Information Acquisition, pp. 378–383 (2005)

21. Lin, H., Lu, H., Zhang, L.: A new automatic recognition system of gender, age and ethnicity. In: Sixth World Congress on Intelligent Control and Automation, vol. 2, pp. 9988–9991 (2002)
22. Hosoi, S., Takikawa, E., Kawade, M.: Ethnicity estimation with facial images. In: Sixth IEEE International Conference on Automatic Face and Gesture Recognition, pp. 195–200 (2004)
23. Ng, C., Tay, Y., Goi, B.: Vision-based human gender recognition: a survey. CoRR (2012), abs/1204.1611
24. Gray, D., Yu, K., Xu, W., Gong, Y.: Predicting facial beauty without landmarks. In: European Conference on Computer Vision (ECCV) (2010)
25. LeCun, Y., Bottou, L., Bengio, Y., Haffner, P.: Gradient-based learning applied to document recognition. Proc. IEEE **86**(11), 2278–2324 (1998)
26. Fu, Y., Guo, G., Huang, T.S.: Age synthesis and estimation via faces: a survey. In: IEEE Transaction on PAMI (2010)
27. Gutta, S., Huang, J., Phillips, P.J., Wechsler, H.: Mixture of experts for classification of gender, ethnic origin, and pose of human faces. In: IEEE Transaction on Neural Networks. Learning System (2000)
28. Lin, C., Wang, S.: Fuzzy support vector machines. IEEE Trans. Neural Netw. **13**(2), 464–471 (2002)
29. Moghaddam, B., Yang, M.: Learning gender with support faces. IEEE Trans. Pattern Anal. Mach. Intell. **24**(5), 707–711 (2002)
30. Zhou, Y., Gu, L., Zhang, H.: Bayesian tangent shape model: estimating shape and pose parameters via Bayesian inference. In: IEEE Computer Society Conference on Computer Vision and Pattern Recognition, vol. 1 (2003)
31. Eisenthal, Y., Dror, G., Ruppin, E.: Facial attractiveness: beauty and the machine. Neural Comput. **18**(1), 119–142 (2006)
32. Kagian, A., Dror, G., Leyvand, T., Cohen-Or, D., Ruppin, E.: A humanlike predictor of facial attractiveness. In: Advances in Neural Information Processing Systems, pp. 649–656 (2005)
33. Likert, R.: Technique for the measurement of attitudes. Arch. Psychol. **22**, 55 (1932)
34. Oliva, A., Torralba, A.: Modeling the shape of the scene: a holistic representation of the spatial envelope. Int. J. Comput. Vis. **42**, 145–175 (2001)
35. Chopra, S., Hadsell, R., LeCun, Y.: Learning a similarity metric discriminatively, with application to face verification. In: IEEE Computer Society Conference on Computer Vision and Pattern Recognition, vol. 1 (2005)
36. Fukushima, K.: Neocognitron: a hierarchical neural network capable of visual pattern recognition. Neural Netw. **1**, 119–130 (1988)
37. Hubel, D., Wiesel, T.: Receptive fields, binocular interaction and functional architecture in the cats visual cortex. J. Physiol. **160**, 106–154 (1962)
38. Huang, G., Jain, V., Amherst, M., Learned-Miller, E.: Unsupervised joint alignment of complex images. In: IEEE International Conference on Computer Vision (2007)
39. Gunes, H., Piccardi, M., Jan, T.: Comparative beauty classification for pre-surgery planning. In: IEEE International Conference on Systems, Man and Cybernetics, vol. 3 (2004)
40. Joy, K., Primeaux, D.: A comparison of two contributive analysis methods applied to an ANN modeling facial attractiveness. In: International Conference on Software Engineering Research, Management and Applications, pp. 82–86 (2006)

Chapter 8
Facial Age Estimation: A Data Representation Perspective

Xin Geng

8.1 Introduction

Age plays a very important role in the human society. People at various ages differ in their behaviors and preferences [2]. As a result, automatic age estimation becomes an indispensable aspect of human-centered computing. Among many age-related traits, face is the most commonly used one for age estimation. As the typical example shown in Fig. 8.1, the appearance of human faces exhibits remarkable changes with the progress of aging, and thus provides relatively reliable information for age estimation. However, since the aging patterns are highly personalized due to various factors including gene, health, lifestyle, weather conditions, etc., facial age estimation by human beings is usually not as accurate as other kinds of facial information, such as identity, expression, and gender. Hence developing automatic facial age estimation methods that are comparable or even superior to the human ability in age estimation has become an attractive yet challenging topic emerging in recent years [8].

For most existing age estimation methods, the problem is formulated as a mapping from the face image x to the age y. It was viewed as either a classification problem or a regression problem [8]. The proposed methods usually focus on the feature extraction from x, the classification/regression process over the $x \rightarrow y$ mapping, or both. For example, in the early work of Lanitis et al. [20, 21], the problem was solved via different regression strategies on a quadratic function called aging function, resulting in the Weighted Appearance Specific method (WAS) and the Appearance and Age Specific method (AAS), respectively. Later, Fu et al. [9, 10] proposed an age estimation method based on multiple linear regression on the discriminative aging manifold of face images. Guo et al. [15] used the SVR (Support Vector Regression) method to design a locally adjusted robust regressor for the prediction of human ages. They later proposed to use the Biologically Inspired Features (BIF) [17] and the

X. Geng (✉)
School of Computer Science and Engineering, Southeast University, 2, Southeast University Road, Nanjing 211189, Jiangsu, China
e-mail: xgeng@seu.edu.cn

Y. Fu (ed.), *Human-Centered Social Media Analytics*,
DOI: 10.1007/978-3-319-05491-9_8, © Springer International Publishing Switzerland 2014

Fig. 8.1 Aging faces of one subject in the FG-NET database [21]. The chronological ages are given under the images

Kernel Partial Least Squares (KPLS) regression [16] for age estimation. Yan et al. [34] regarded age estimation as a regression problem with non-negative label intervals and solved the problem through semidefinite programming. They also proposed an EM algorithm to solve the regression problem and speed up the optimization process [33]. By using the Spatially Flexible Patch (SFP) as the feature descriptor, the age regression was further improved with the patch-based Gaussian mixture model [36] and the patch-based hidden Markov model [38]. Noticing the advantages of personalized age estimation, Zhang and Yeung [37] formulated the problem as a multitask learning problem and proposed the multitask warped Gaussian process to learn a separate age estimator for each person. In order to build a robust facial age estimation system, Ni et al. [23, 24] proposed a method based on the mining of the noisy aging face images collected from the Web images and videos. One of the most recent progresses was made by Chang et al. [3], who transformed an age estimation task into multiple cost-sensitive binary classification subproblems, and solved the problem with an ordinal hyperplane ranking algorithm.

Although many effective age estimation methods have been proposed with assumption of the basic data representation form $x \rightarrow y$, other options might be worthwhile to consider due to the interesting characteristics of the age estimation problem. Compared with other facial variations, the aging effects exhibit at least the following characteristics:

1. *Personalized aging patterns.* Different people age in different ways. The aging pattern of each person is determined by his/her genes as well as many external factors, such as health, lifestyle, weather conditions, *etc.*
2. *Aging patterns are temporal.* The aging progress must obey the order of time. The face status at a particular age will affect all older faces, but will not affect those younger ones.
3. *Aging progress is gradual.* The facial appearance changes gradually with the growth of age. As a result, faces at close ages look quite similar while those with bigger age difference vary more apparently.

4. *Aging progress is uncontrollable.* No one can advance or delay aging at will. The procedure of aging is slow and irreversible.

Characteristic 1 indicates that the aging pattern model should be personalized, i.e., the x in the $x \rightarrow y$ mapping should be somehow organized according to the personal identity. Characteristic 2 determines that the facial age y in the $x \rightarrow y$ mapping is totally ordered. Each age has a unique rank in the time sequence. Characteristic 3 makes it possible to utilize the face images at the neighboring ages while learning a particular age since they are quite similar. One possible solution to all the above issues is to use new data representation forms (and usually with machine learning methods matching these data representation forms) specially designed for the age estimation problem. Finally, characteristic 4 implies that the collection of training data for age estimation is extremely laborious, which usually requires great effort in searching for photos taken years ago. As a result, the aging data can hardly be sufficient and complete. The available datasets [21, 27] typically just contain a very limited number of aging images for each person, and the images at the higher ages are especially rare. No matter what data representation is used, insufficient and incomplete training set is always a problem likely to face up to in the age estimation problem.

This chapter presents two typical solutions to facial age estimation which are based on special data representation forms other than the traditional $x \rightarrow y$ mapping. The first is called AGES (AGing pattErn Subspace) [13, 14]. The basic idea of AGES is to model the aging pattern, which is defined as the sequence of a particular individual's face images sorted in time order, by constructing a representative subspace. The proper aging pattern for a previously unseen face image is determined by the projection in the subspace that can reconstruct the face image with minimum reconstruction error, while the position of the face image in that aging pattern will then indicate its age. The second solution is based a new learning paradigm named *label distribution learning* [11, 12]. The basic idea is to regard each face image as an instance associated with a label distribution. The label distribution covers a certain number of class labels, representing the degree that each label describes the instance. Through this way, one face image can contribute to not only the learning of its chronological age, but also the learning of its adjacent ages.

The rest of this chapter is organized as follows: Section 8.2 introduces the AGES method for facial age estimation. Section 8.3 shows how to apply label distribution learning to facial age estimation. Section 8.4 reports some experimental results of the aforementioned methods. Finally, Sect. 8.5 summarize this chapter.

8.2 Facial Age Estimation Based on AGing pattErn Subspace (AGES)

As mentioned in Sect. 8.1, the aging progress is personal and temporal. From the data representation point of view, these characteristics can hardly be reflected by the $x \rightarrow y$ mapping. Some existing work [3, 37] relies on the feature extraction step or the classification/regression step to deal with such characteristics. However, a good data representation matching these characteristics might greatly reduce the dependency on those sophisticated feature extractors, classifiers, or regressors. This section introduces one such solution to facial age estimation. It is based on a specific data representation named *aging pattern*, which naturally incorporates the concept of personal identity and time. Since all the labels are implicitly included in the data representation, the originally supervised $x \rightarrow y$ mapping is transformed into an unsupervised process.

8.2.1 Aging Pattern

A formal definition of aging pattern is given as follows:

Definition 1. An aging pattern is a sequence of personal face images sorted in time order.

The keywords in this definition are "personal" and "time." All face images in an aging pattern must come from the same person, and they must be ordered by time. Suppose a gray-scale face image is represented by a two-dimensional matrix $\mathbf{I}$, where $\mathbf{I}(x, y)$ is the intensity of the pixel (x, y), then the aging pattern of a person can be represented by a three-dimensional matrix $\mathbf{P}$, where $\mathbf{P}(x, y, t)$ is the intensity of the pixel (x, y) in his/her face image at the time t. In this chapter, age is regarded as a non-negative integer. So, the value of t can only be a non-negative integer. Take the aging pattern shown in Fig. 8.2 as an example. Along the t axis, each age (from 0 to 8 in this example) is allocated one position. If face images are available for certain ages (in this case 2, 5, and 8), they are filled into the corresponding positions. If not, the positions are left blank (dotted squares). If all positions are filled, the aging pattern is called a *complete aging pattern*, otherwise it is called an *incomplete aging pattern*.

Before the aging pattern can be further processed, the face images in it are first transformed into feature vectors. This can be done by any face image feature extractors, such as the well-known Appearance Model [6]. Suppose the feature vector $\boldsymbol{b}$ extracted by the Appearance Model is n-dimensional, the number of interested ages is p. Then the aging pattern can be represented by an $(n \times p)$-dimensional feature vector $\boldsymbol{x}$. Each age is allocated n consecutive elements in $\boldsymbol{x}$. If the interested ages are sorted in ascending order, then the k-th age occupies from the $((k-1) \times n + 1)$-th element to the $(k \times n)$-th element in $\boldsymbol{x}$. If the face image at a particular age is not available, then the corresponding part in $\boldsymbol{x}$ is marked as missing features. Figure 8.2 gives an example of the vectorization of the aging pattern when the interested ages are from 0 to 8. $\boldsymbol{b}_2$, $\boldsymbol{b}_5$ and $\boldsymbol{b}_8$ represent the feature vectors of the face images at the ages 2, 5 and 8, respectively.

By representing aging patterns in this way, the age and personal ID are naturally integrated into the data without any preassumptions. Each aging pattern implies one ID, each age is fixed into a position in the aging pattern, and the position is ordered

Fig. 8.2 Vectorization of the aging pattern. The ages (0–8) are marked at the *top-left* to the corresponding positions and above the corresponding feature vectors. The missing parts in the aging pattern vector are marked by 'm'

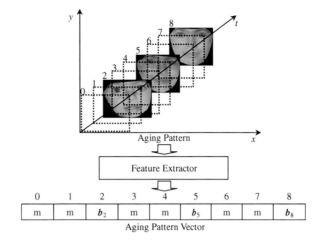

according to time. Consequently, Characteristic 1 and 2 mentioned in Sect. 8.1 can be well utilized. So long as the aging patterns are well sampled, i.e., the training aging patterns are sufficient enough to represent the real distribution of all aging patterns, a proper model of aging pattern can be learned for age estimation. Since all the age labels are implied in the data representation, the learning process becomes unsupervised. However, this brings two other challenges: (1) During training, the learning algorithm applied to the aging patterns must be able to handle highly incomplete training samples; (2) During age estimation, the most suitable aging pattern as well as the most suitable position in that aging pattern must be selected for an unknown face image. The next section mainly tackles these two problems.

8.2.2 The AGES Algorithm

8.2.2.1 Subspace Learning on Aging Patterns

As mentioned before, since all the class labels are implied in the data representation, the modeling of aging pattern becomes an unsupervised learning process. A representative model for the aging patterns can be built up by the information theory approach of coding and decoding. For example, we can use PCA [19] to construct a subspace that captures as much variation as possible in the dataset. The projection in the subspace is computed by

$$y = \mathbf{W}^{\mathrm{T}}(x - \mu), \tag{8.1}$$

where μ is the mean vector of x, and $\mathbf{W}^{\mathrm{T}}$ is the transpose of $\mathbf{W}$, which is composed by the eigenvectors of the covariance matrix. The problem is that the aging pattern vector x is highly incomplete. There exist several approaches to PCA with missing data [19]. However, the statistical distribution of the aging patterns is unlikely to be normal, thus those methods based on the assumption of normal distribution [28, 30] are not suitable for this problem. Moreover, the aging pattern vector is highly incomplete, thus those methods dealing with minor missing data [28, 32] are also not suitable. Based on the characteristics of aging patterns, an EM-like iterative learning algorithm is proposed here to learn a representative subspace.

Suppose the training set has N aging pattern vectors $D = \{x_1, \ldots, x_N\}$. Any instance in this set can be written as $x_k = \{x_k^a, x_k^m\}$, where x_k^a is the available features and x_k^m is the missing features of x_k. Once the transformation matrix $\mathbf{W}$ is determined, the projection y_k of x_k in the subspace can be calculated by Eq. (8.1), and the reconstruction of x_k is calculated by

$$\hat{x}_k = \mu + \mathbf{W} y_k. \tag{8.2}$$

$\hat{x}_k$ can also be written as $\hat{x}_k = \{\hat{x}_k^a, \hat{x}_k^m\}$, where $\hat{x}_k^a$ is the reconstruction of x_k^a, and $\hat{x}_k^m$ is the reconstruction of x_k^m. Then the mean reconstruction error (residuals) of the dataset D in the subspace spanned by $\mathbf{W}$ is calculated by

$$\bar{\varepsilon} = \frac{1}{N} \sum_{k=1}^{N} (x_k - \hat{x}_k)^{\mathrm{T}} (x_k - \hat{x}_k). \tag{8.3}$$

It is well known that standard PCA can be derived by minimizing $\bar{\varepsilon}$ [19]. With the presence of the missing features x_k^m, the goal is changed into finding a $\mathbf{W}$ that minimizes the mean reconstruction error of the available features

$$\bar{\varepsilon}^a = \frac{1}{N} \sum_{k=1}^{N} (x_k^a - \hat{x}_k^a)^{\mathrm{T}} (x_k^a - \hat{x}_k^a). \tag{8.4}$$

When initializing, x_k^m is replaced by the mean vector $[\mu(_k^m)]$, calculated from the samples for which the features at the same positions as x_k^m are available. Then standard PCA is applied to the full-filled dataset to get the initial transformation matrix $\mathbf{W}_0$ and mean vector μ_0. In the iteration i, the projection of x_k in the subspace spanned by $\mathbf{W}_i$ is estimated first. Since there are many missing features in x_k, the projection cannot be computed directly by Eq. (8.1). Note that the aging patterns are highly redundant, it is possible to estimate y_k only based on part of x_k [22], say x_k^a. Instead of using inner product, y_k is solved as the least squares solution of

$$[\mathbf{W}_i(_k^a)] y_k = x_k^a - [\mu_i(_k^a)], \tag{8.5}$$

where $[\mathbf{W}_i(_k^a)]$ is the part in $\mathbf{W}_i$ and $[\mu_i(_k^a)]$ is the part in μ_i that correspond to the positions of x_k^a. Suppose the subspace is d-dimensional, then there are d unknowns in y_k. Thus at least d available features (elements in x_k^a) are needed to over-constrain the problem, but more may be required due to linear dependencies among the equations. In practice, 2–3 times as many available features as unknowns are typically required to get a reasonable solution [22]. In the experiments described later, most configurations satisfy the condition $L(x_k^a) > 3d$, where $L(x_k^a)$ is the number of elements in x_k^a. After getting the estimation of y_k, $\hat{x}_k$ is calculated by Eq. (8.2) and x_k^m is updated by $\hat{x}_k^m$. Then, standard PCA is applied to the updated dataset to get the new transformation matrix $\mathbf{W}_{i+1}$ and mean vector μ_{i+1}. The whole process repeats until the maximum iteration τ is exceeded or $\bar{\varepsilon}^a$ is smaller than a predefined threshold ε. The pseudocode of the training process of AGES is shown at the left side of Table 8.1. The convergence of this algorithm is proved as follows:

Proof. Suppose in iteration i, the training data is $x_k^{(i)}$, the reconstruction of $x_k^{(i)}$ by $\mathbf{W}_i$ is $[\hat{x}_k^{(i)}(\mathbf{W}_i)]$, the reconstruction error of $x_k^{(i)}$ by $\mathbf{W}_i$ is $\varepsilon(x_k^{(i)}, \mathbf{W}_i)$, and the reconstruction error of the available features is $\varepsilon^a(x_k^{(i)}, \mathbf{W}_i)$. Note that $[\hat{x}_k^{(i)}(\mathbf{W}_i)]$ and the data of the next iteration, $x_k^{(i+1)}$, share the same values at the positions of missing features, so

$$\varepsilon^a(x_k^{(i)}, \mathbf{W}_i) = U\left([\hat{x}_k^{(i)}(\mathbf{W}_i)], x_k^{(i+1)}\right),\tag{8.6}$$

where $U(v_1, v_2)$ denotes the squared Euclidean distance between v_1 and v_2. Consequently, $\bar{\varepsilon}_i^a = \bar{U}$, where $\bar{\varepsilon}_i^a$ is the $\bar{\varepsilon}^a$ of iteration i and $\bar{U}$ is the mean value of $U([\hat{x}_k^{(i)}(\mathbf{W}_i)], x_k^{(i+1)})$. If $x_k^{(i+1)}$ is also reconstructed by $\mathbf{W}_i$, then

$$\begin{aligned}\varepsilon(x_k^{(i+1)}, \mathbf{W}_i) &= U\left([\hat{x}_k^{(i+1)}(\mathbf{W}_i)], x_k^{(i+1)}\right)\\ &\leq U\left([\hat{x}_k^{(i)}(\mathbf{W}_i)], x_k^{(i+1)}\right)\end{aligned}\tag{8.7}$$

because the line between $[\hat{x}_k^{(i+1)}(\mathbf{W}_i)]$ and $x_k^{(i+1)}$ is orthogonal to the subspace spanned by $\mathbf{W}_i$ so that they have the minimum Euclidean distance. Consequently, $\bar{\varepsilon}(x_k^{(i+1)}, \mathbf{W}_i) \leq \bar{U}$, where $\bar{\varepsilon}(x_k^{(i+1)}, \mathbf{W}_i)$ is the mean reconstruction error of $x_k^{(i+1)}$ by $\mathbf{W}_i$. After applying PCA on $x_k^{(i+1)}$, the new transformation matrix $\mathbf{W}_{i+1}$ minimizes the mean reconstruction error, thus

$$\bar{\varepsilon}(x_k^{(i+1)}, \mathbf{W}_{i+1}) \leq \bar{\varepsilon}(x_k^{(i+1)}, \mathbf{W}_i).\tag{8.8}$$

Obviously, $\bar{\varepsilon}_{i+1}^a \leq \bar{\varepsilon}(x_k^{(i+1)}, \mathbf{W}_{i+1})$. So

$$\bar{\varepsilon}_{i+1}^a \leq \bar{\varepsilon}(x_k^{(i+1)}, \mathbf{W}_{i+1}) \leq \bar{\varepsilon}(x_k^{(i+1)}, \mathbf{W}_i) \leq \bar{U} = \bar{\varepsilon}_i^a.\tag{8.9}$$

Thus the algorithm will converge to minimize $\bar{\varepsilon}^a$.

During the training process of AGES, the missing faces in the training aging patterns can be simultaneously learned, as a "byproduct," by reconstructing the whole

Table 8.1 Pseudocode of the AGES algorithm

Training process	Test process
Input: $D = \{x_1, \ldots, x_N\}$	**Input:** b, **W**, μ
Output: **W**, μ	**Output:** age
$i \leftarrow 0$; $x_k^m \leftarrow [\mu(_k^{(m)})]$;	**for** $j \leftarrow 1$ **to** p **do**
Apply PCA to the full-filled D to get $\mathbf{W}_0$;	Place b in the j-th position of z_j;
$\mu_0 \leftarrow$ the mean vector of the full-filled D;	Estimate y_j by Eq. (8.10); Calculate $\varepsilon^a(j)$ by Eq. (8.11);
while $i < \tau$ **and** $\bar\varepsilon^a > \varepsilon$	**end**
Estimate y_k by Eq. (8.5);	$r \leftarrow \arg\min_j(\varepsilon^a(j))$;
Reconstruct x_k by Eq. (8.2);	$age \leftarrow$ the age associated to the r-th position;
$x_k^m \leftarrow \hat{x}_k^m$;	
Apply PCA to the updated D to get $\mathbf{W}_{i+1}$;	
$\mu_{i+1} \leftarrow$ the mean vector of the updated D;	
$i \leftarrow i + 1$;	
end	
$\mathbf{W} \leftarrow \mathbf{W}_i$; $\mu \leftarrow \mu_i$;	

aging pattern vectors through Eq. (8.2). Figure 8.3 shows some typical examples of the "full-filled" aging patterns when AGES is applied to the FG-NET Aging Database [7]. For clarity, only the faces in the most changeable age range from 0 to 18 with 2 years as interval are shown. It can be seen that the learned faces inosculate with those real faces very well in the aging patterns. Thus this learning algorithm can also be used to simulate aging effects on human faces. Note that the learned faces are generated via an inverse process of the Appearance Model. So, it is possible for them to have different poses due to different reconstructed shape models. It is also noteworthy that there is no significant ghosting or blurring in the learned faces, which is common in face synthesis through subspace learning algorithms. Moreover, in some cases, such as the second line of Fig. 8.3, the learned faces might be the majority in the aging pattern. These prove the ability of the proposed algorithm to learn from highly incomplete data.

The process of the learning algorithm is actually a process of interaction between the global aging pattern model and the personalized aging patterns. Although different persons age in different ways, the commonality of all aging patterns (the general trend of aging) modeled by the subspace is also crucial for age estimation, especially when the aging patterns are highly incomplete. In each iteration, the missing part of the personal aging pattern is first estimated by the current global aging pattern model. Then, the global model is further refined by the updated personal aging patterns. In this way, the commonality and the personality of the aging patterns are alternately utilized to learn the final subspace.

0	2	4	6	8	10	12	14	16	18

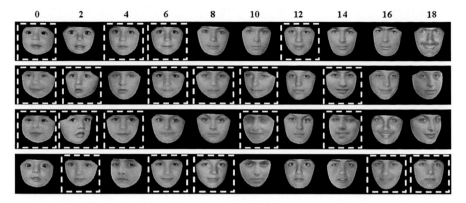

Fig. 8.3 The 'full-filled' aging patterns. Each *line* shows the aging pattern of one person from the age 0 to 18. The ages are marked above the corresponding faces. The faces learned by the algorithm are surrounded by the *dashed squares*

8.2.2.2 Age Estimation

The aging pattern subspace is a global model for aging patterns, each of which corresponds to a sequence of age labels. But the task of age estimation is usually based on a single face image input, and expects a single age output. This section will describe how this can be done based on the aging pattern subspace.

Given a previously unseen face image $\mathbf{I}$, its feature vector b is first extracted by the feature extractor. According to Characteristic 1 and 2 mentioned in Sect. 8.1, age estimation should involve at least two main steps: Step 1 is to determine the suitable aging pattern for a particular face and Step 2 is to find the position of the face in that aging pattern. Note that each point in the aging pattern subspace corresponds to one aging pattern. Thus in Step 1, the proper aging pattern for $\mathbf{I}$ can be selected through finding a point in the aging pattern subspace that can best reconstruct b, i.e. minimize the reconstruction error. However, without knowing the position of $\mathbf{I}$ in the aging pattern, which should be determined in the second step, the reconstruction error cannot be actually calculated. Thus $\mathbf{I}$ is placed at every possible position in the aging pattern, getting p aging pattern vectors $z_j (j = 1 \ldots p)$ by placing b at the position j in z_j. For each z_j, the projection with minimum reconstruction error, y_j, can be estimated by the least squares solution of

$$\mathbf{W}_{(j)} y_j = b - \mu_{(j)} \tag{8.10}$$

where $\mu_{(j)}$ is the part in μ and $\mathbf{W}_{(j)}$ is the part in $\mathbf{W}$ that correspond to the position j. The reconstruction error of y_j can be calculated by

$$\varepsilon^a(j) = (b - \mu_{(j)} - \mathbf{W}_{(j)} y_j)^{\mathrm{T}} (b - \mu_{(j)} - \mathbf{W}_{(j)} y_j), \tag{8.11}$$

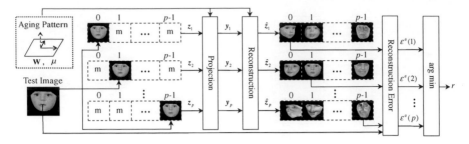

Fig. 8.4 Flowchart of age estimation in AGES. The missing parts in the aging patterns are marked by "m." The test image is from a 1-year-old baby and the interested ages are from 0 to $p-1$

Then, the projection y_r that can reconstruct b with minimum reconstruction error over all the p possible positions is determined by

$$r = \arg\min_{j}(\varepsilon^a(j)). \tag{8.12}$$

Thus the suitable aging pattern for $\mathbf{I}$ is z_r. Step 2 afterward becomes trivial because r also indicates the position of $\mathbf{I}$ in z_r. Finally, the age associated to the position r is returned as the estimated age of $\mathbf{I}$. The pseudocode of the age estimation process (test process) in AGES is shown at the right side in Table 8.1.

The flowchart of the age estimation process is shown in Fig. 8.4. To make it more understandable, the face images (both the original and reconstructed ones) instead of feature vectors are shown in the aging patterns. It is interesting to note that when the test image is placed at a wrong position in the aging pattern, such as placing the 1-year-old face at the position for the $(p-1)$-year-old face, the reconstructed faces become ghost-like twisted faces. On the other hand, if the test image is placed at the right position, the aging pattern subspace can not only reconstruct the original face very well, but also reasonably derive all other faces in the aging pattern. This vividly proves the ability of the aging pattern subspace to distinguish different facial ages. Moreover, as a byproduct of age estimation, the faces at different ages of the subject in the test image can be simulated at the same time without additional computation.

During the age estimation process of AGES, the proper aging pattern for the test image is generated based on both the aging pattern subspace and the face image feature. The subspace defines the general trend of aging, and the face image feature represents the personalized factors. By placing the feature vector at different positions, the candidate aging patterns specified to the test face are generated. Among these candidates, only one is consistent with the general aging trend, which can be detected via minimum reconstruction error by the aging pattern subspace. At the same time, the position of the test image in that aging pattern can be determined.

8.3 Facial Age Estimation Based on Label Distribution Learning

As mentioned in Sect. 8.1, since the aging progress is slow and irreversible, the available training data for facial age estimation can hardly be sufficient and complete. How to make the best use of the limited training data thus becomes an important issue of the facial age estimation problem. Fortunately, the aging progress is also gradual, which means that the faces at close ages look quite similar while those with bigger age difference look more different from each other. This motivates us to use the neighboring ages while learning a particular age. The idea is well beyond the routine of the $x \rightarrow y$ mapping. In order to utilize the neighboring ages, a face image x is labeled by not only its chronological age y, but also those ages close to y. Of course, the further an age is away from y, the less it describes the face. The description degree of an age to the face is indicated by a real number within $[0, 1]$. In this way, a face image is labeled by a new supervision information called *label distribution D*, and the $x \rightarrow y$ mapping is transformed into the $x \rightarrow D$ mapping.

8.3.1 Label Distribution

In the label distribution of an instance x, a real number $d_x^y \in [0, 1]$ called *description degree* is assigned to each label y, representing the degree that y describes x. The description degrees of all the labels sum up to 1, indicating a full class description of the instance. Since age is essentially a continuous time spectrum, the age label distribution can be defined as a continuous distribution. But in practice, age is usually measured in years, which is actually a discrete sampling over the time spectrum. Thus the label distribution is defined as a discrete distribution in this chapter. Under this definition, the traditional ways to label an instance with a single label or multiple labels can all be viewed as special cases of label distribution. Some typical examples of the label distributions for five class labels are shown in Fig. 8.5. For case (a), a single label is assigned to the instance, so $d_x^{y_2} = 1$ means that the class label y_2 fully describes the instance. For case (b), two labels (y_2 and y_4) are assigned to the instance, so each of them by default describes 50% of the instance, i.e., $d_x^{y_2} = d_x^{y_4} = 0.5$. Finally, case (c) represents a general case of label distribution satisfying the constraints $d_x^y \in [0, 1]$ and $\sum_y d_x^y = 1$.

Special attention should be paid to the meaning of d_x^y, which is *not* the *probability* that y correctly labels x, but the proportion that y accounts for in a full class description of x. Thus, all the labels with a nonzero description degree are actually the "correct" labels to describe the instance, but just with different importance measured by d_x^y. Recognizing this, one can distinguish label distribution from the previous studies on probabilistic labels [5, 26, 29], where the basic assumption is that there is only one 'correct' label for each instance. Probabilistic labels are mainly used in the cases where the real label of the instance cannot be obtained with certainty.

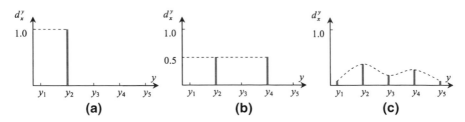

Fig. 8.5 Three cases of label distribution: **a** single label, **b** multiple labels, and **c** a general case of label distribution

In practice, it is usually difficult to determine the probability (or confidence) of a label. In most cases, it relies on the prior knowledge of the human experts, which is a highly subjective and variable process. As a result, the problem of learning from probabilistic labels has not been extensively studied to date. Fortunately, although not a probability by definition, d_x^y still shares the same constraints with probability, i.e., $d_x^y \in [0, 1]$ and $\sum_y d_x^y = 1$. Thus many theories and methods in statistics can be applied to label distributions.

It is also worthwhile to distinguish description degree from the concept *membership* used in *fuzzy classification* [39]. Membership is a truth value that may range between completely true and completely false. It is designed to handle the status of *partial truth* which often appears in the non-numeric linguistic variables. For example, the age 25 might have a membership of 0.7 to the linguistic category "young," and 0.3 to "middle age." But for a particular face, its association with the chronological age 25 will be either completely true or completely false. On the other hand, description degree reflects the *ambiguity* of the class description of the instance, i.e., one class label may only partially describe the instance. For example, due to the appearance similarity of the neighboring ages, both the chronological age 25 and the neighboring ages 24 and 26 can be used to describe the appearance of a 25-year-old face. For each of 24, 25, and 26, it is completely true that it can be used to describe the face (in the sense of appearance). Each age's description degree indicates how much the age contributes to the full class description of the face.

The prior label distribution assigned to a face image at the chronological age α should satisfy the following two properties: (1) The description degree of α is the highest in the label distribution, which ensures the leading position of the chronological age in the class description; (2) The description degree of other ages decreases with the increase of the distance away from α, which makes the age closer to the chronological age contribute more to the class description. While there are many possibilities, Fig. 8.6 shows two kinds of prior label distributions for the images at the chronological age α, i.e., the Gaussian distribution and the triangle distribution. Note that the age y is regarded as a discrete class label in this chapter while both the Gaussian and triangle distributions are defined by continuous density functions $p(y)$. Directly letting $d_x^y = p(y)$ might induce $\sum_y d_x^y \neq 1$. Thus a normalization process $d_x^y = p(y)/\sum_y p(y)$ is required to ensure $\sum_y d_x^y = 1$.

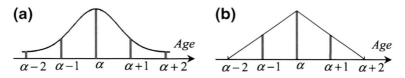

Fig. 8.6 Typical label distributions for the image at the chronological age α: **a** Gaussian distribution and **b** triangle distribution

8.3.2 Learning from Label Distributions

8.3.2.1 Problem Formulation

As mentioned before, many theories and methods from statistics can be borrowed to deal with label distributions. First of all, the description degree d_x^y could be represented by the form of conditional probability, i.e., $d_x^y = P(y|x)$. This might be explained as that given an instance x, the probability of the presence of y is equal to its description degree. Then, the problem of label distribution learning can be formulated as follows:

Let $\mathscr{X} = \mathbb{R}^q$ denote the input space and $\mathscr{Y} = \{y_1, y_2, \ldots, y_c\}$ denote the finite set of possible class labels. Given a training set $S = \{(x_1, D_1), (x_2, D_2), \ldots, (x_n, D_n)\}$, where $x_i \in \mathscr{X}$ is an instance, $D_i = \{d_{x_i}^{y_1}, d_{x_i}^{y_2}, \ldots, d_{x_i}^{y_c}\}$ is the label distribution associated with x_i, the goal of label distribution learning is to learn a conditional probability mass function $p(y|x)$ from S, where $x \in \mathscr{X}$ and $y \in \mathscr{Y}$.

For the problem of age estimation, suppose the same shape of prior label distribution (e.g., Fig. 8.6) is assigned to each face image, then the highest description degree for each image will be the same, say, p_{max}. Since the description degree of the chronological age should always be the highest in the label distribution, for a face image x_α at the chronological age α, the label distribution learner should output

$$p(\alpha|x_\alpha) = p_{max}, \tag{8.13}$$
$$p(\alpha + \Delta|x_\alpha) = p_{max} - p_\Delta, \tag{8.14}$$

where $p_\Delta \in [0, 1]$ is the description degree difference from p_{max} when the age changes to a neighboring age $\alpha + \Delta$. Similarly, for a face image $x_{\alpha + \Delta}$ at the chronological age $\alpha + \Delta$,

$$p(\alpha + \Delta|x_{\alpha + \Delta}) = p_{max}. \tag{8.15}$$

As mentioned before, the faces at the close ages are quite similar, i.e., $x_{\alpha+\Delta} \approx x_\alpha$. Suppose the feature space is continuous, then,

$$p(\alpha + \Delta|x_{\alpha+\Delta}) \approx p(\alpha + \Delta|x_\alpha). \tag{8.16}$$

So, p_Δ is a small positive number, which indicates that $p(\alpha + \Delta | x_\alpha)$ is just a little bit smaller than $p(\alpha | x_\alpha)$. Note that the above analysis does not depend on any particular form of the prior label distribution except that it must satisfy the two properties mentioned in Sect. 8.3.1. This proves that when applied to age estimation, label distribution learning tends to learn the similarity among the neighboring ages, no matter what the (reasonable) prior label distribution might be.

Suppose $p(y|x)$ is a parametric model $p(y|x; \theta)$, where θ is the vector of the model parameters. Given the training set S, the goal of label distribution learning is to find the θ that can generate a distribution similar to D_i given the instance x_i. If the Kullback-Leibler divergence is used as the measurement of the similarity between two distributions, then the best model parameter vector θ^* is determined by:

$$\theta^* = \operatorname*{argmin}_{\theta} \sum_i \sum_j \left(d_{x_i}^{y_j} \ln \frac{d_{x_i}^{y_j}}{p(y_j|x_i; \theta)} \right)$$

$$= \operatorname*{argmax}_{\theta} \sum_i \sum_j d_{x_i}^{y_j} \ln p(y_j|x_i; \theta). \tag{8.17}$$

It is interesting to examine the traditional learning paradigms under the optimization criterion shown in Eq. (8.17). For *single-label learning* (see Fig. 8.5a), $d_{x_i}^{y_j} = Kr(y_j, y(x_i))$, where $Kr(\cdot, \cdot)$ is the Kronecker delta function and $y(x_i)$ is the single class label of x_i. Consequently, Eq. (8.17) can be simplified to

$$\theta^* = \operatorname*{argmax}_{\theta} \sum_i \ln p(y(x_i)|x_i; \theta). \tag{8.18}$$

This is actually the maximum likelihood (ML) estimation of θ. The later use of $p(y|x; \theta)$ for classification is equivalent to the maximum a posteriori (MAP) decision.

For *multilabel learning* [31], each instance is associated with a label set (see Fig. 8.5b). Consequently, Eq. (8.17) can be changed into

$$\theta^* = \operatorname*{argmax}_{\theta} \sum_i \frac{1}{|Y_i|} \sum_{y \in Y_i} \ln p(y|x_i; \theta), \tag{8.19}$$

where Y_i is the label set associated with x_i. Equation (8.19) can be viewed as a ML criterion weighted by the reciprocal cardinality of the label set associated with each instance. In fact, this is equivalent to first applying the Entropy-based Label Assignment (ELA) [31], a well-known technique dealing with multilabel data, to transform the multilabel instances into the weighted single-label instances, and then optimizing the ML criterion based on the weighted single-label instances.

It can be seen from the above analysis that with proper constraints, a label distribution learning model can be transformed into the commonly used methods for single-label or multilabel learning. Thus, label distribution learning is a more general learning framework which includes single-label learning as its special case. Accordingly, the algorithms that learn from the label distributions should be designed within this new learning framework.

8.3.2.2 The Learning Process

Suppose $f_k(x, y)$ is a *feature function* which depends on both the instance x and the label y. Then, the expected value of f_k w.r.t. the empirical joint distribution $\tilde{p}(x, y)$ in the training set is

$$\tilde{f}_k = \sum_y \int \tilde{p}(x, y) f_k(x, y) \, dx. \tag{8.20}$$

The expected value of f_k w.r.t. the conditional model $p(y|x; \boldsymbol{\theta})$ and the empirical distribution $\tilde{p}(x)$ in the training set is

$$\hat{f}_k = \sum_y \int \tilde{p}(x) p(y|x; \boldsymbol{\theta}) f_k(x, y) \, dx. \tag{8.21}$$

One reasonable choice of $p(y|x; \boldsymbol{\theta})$ is the one that has the maximum *conditional entropy* subject to the constraint $\tilde{f}_k = \hat{f}_k$. It can be proved [1] that such a model (a.k.a. the *maximum entropy model*) has the exponential form

$$p(y|x; \boldsymbol{\theta}) = \frac{1}{Z} \exp\left(\sum_k \theta_k f_k(x, y)\right), \tag{8.22}$$

where $Z = \sum_y \exp\left(\sum_k \theta_k f_k(x, y)\right)$ is the normalization factor and θ_k is the k-th model parameter in $\boldsymbol{\theta}$. In practice, the features usually depend only on the instance but not on the class label. Thus, Eq. (8.22) can be rewritten as

$$p(y|x; \boldsymbol{\theta}) = \frac{1}{Z} \exp\left(\sum_k \theta_{y,k} g_k(x)\right), \tag{8.23}$$

where $g_k(x)$ is a class-independent feature function.

Substituting Eq. (8.23) into Eq. (8.17) and recognizing $\sum_j d_{x_i}^{y_j} = 1$ yields the target function of $\boldsymbol{\theta}$

$$T(\boldsymbol{\theta}) = \sum_{i,j} d_{\boldsymbol{x}_i}^{y_j} \ln p(y_j | \boldsymbol{x}_i; \boldsymbol{\theta})$$

$$= \sum_{i,j} d_{\boldsymbol{x}_i}^{y_j} \sum_k \theta_{y_j,k} g_k(\boldsymbol{x}_i)$$

$$- \sum_i \ln \sum_j \exp\left(\sum_k \theta_{y_j,k} g_k(\boldsymbol{x}_i) \right). \tag{8.24}$$

Directly setting the gradient of Eq. (8.24) w.r.t. $\boldsymbol{\theta}$ to zero does not yield a closed-form solution. Thus the optimization of Eq. (8.24) uses a strategy similar to Improved Iterative Scaling (IIS) [25], a well-known algorithm for maximizing the likelihood of the maximum entropy model. IIS starts with an arbitrary set of parameters. Then for each step, it updates the current estimate of the parameters $\boldsymbol{\theta}$ to $\boldsymbol{\theta} + \boldsymbol{\Delta}$, where $\boldsymbol{\Delta}$ maximizes a lower bound to the change in likelihood $\Omega = T(\boldsymbol{\theta} + \boldsymbol{\Delta}) - T(\boldsymbol{\theta})$. This iterative process, nevertheless, needs to be migrated to the new target function $T(\boldsymbol{\theta})$. Furthermore, the constraint on the feature functions required by IIS, $f_k(\boldsymbol{x}, y) \geq 0$ (hence $g_k(\boldsymbol{x}) \geq 0$) should be removed to ensure the freedom in choosing any feature extractors suitable for the data.

In detail, the change of $T(\boldsymbol{\theta})$ between adjacent steps is

$$T(\boldsymbol{\theta} + \boldsymbol{\Delta}) - T(\boldsymbol{\theta}) = \sum_{i,j} d_{\boldsymbol{x}_i}^{y_j} \sum_k \delta_{y_j,k} g_k(\boldsymbol{x}_i)$$

$$- \sum_i \ln \sum_j p(y_j | \boldsymbol{x}_i; \boldsymbol{\theta}) \exp\left(\sum_k \delta_{y_j,k} g_k(\boldsymbol{x}_i) \right), \tag{8.25}$$

where $\delta_{y_j,k}$ is the increment for $\theta_{y_j,k}$. Applying the inequality $-\ln x \geq 1 - x$ yields

$$T(\boldsymbol{\theta} + \boldsymbol{\Delta}) - T(\boldsymbol{\theta}) \geq \sum_{i,j} d_{\boldsymbol{x}_i}^{y_j} \sum_k \delta_{y_j,k} g_k(\boldsymbol{x}_i) + n$$

$$- \sum_{i,j} p(y_j | \boldsymbol{x}_i; \boldsymbol{\theta}) \exp\left(\sum_k \delta_{y_j,k} g_k(\boldsymbol{x}_i) \right). \tag{8.26}$$

Differentiating the right side of Eq. (8.26) w.r.t. $\delta_{y_j,k}$ yields the coupled equations of $\delta_{y,k}$ which are hard to be solved. To decouple the interaction among $\delta_{y,k}$, Jensen's inequality is applied here, i.e., for a probability mass function $p(x)$,

$$\exp\left(\sum_x p(x) q(x) \right) \leq \sum_x p(x) \exp\left(q(x) \right). \tag{8.27}$$

The last term of Eq. (8.26) can be rewritten as:

$$\sum_{i,j} p(y_j|\boldsymbol{x}_i; \boldsymbol{\theta}) \exp\left(\sum_k \delta_{y_j,k} s\left(g_k(\boldsymbol{x}_i)\right) g^\#(\boldsymbol{x}_i) \frac{|g_k(\boldsymbol{x}_i)|}{g^\#(\boldsymbol{x}_i)}\right), \qquad (8.28)$$

where $g^\#(\boldsymbol{x}_i) = \sum_k |g_k(\boldsymbol{x}_i)|$ and $s\left(g_k(\boldsymbol{x}_i)\right)$ is the sign of $g_k(\boldsymbol{x}_i)$. Since $|g_k(\boldsymbol{x}_i)|/g^\#(\boldsymbol{x}_i)$ can be viewed as a probability mass function, Jensen's inequality can be applied to Eq. (8.26) to yield

$$T(\boldsymbol{\theta}+\boldsymbol{\Delta}) - T(\boldsymbol{\theta}) \geq \sum_{i,j} d_{\boldsymbol{x}_i}^{y_j} \sum_k \delta_{y_j,k} g_k(\boldsymbol{x}_i) + n$$
$$- \sum_{i,j} p(y_j|\boldsymbol{x}_i; \boldsymbol{\theta}) \sum_k \frac{|g_k(\boldsymbol{x}_i)|}{g^\#(\boldsymbol{x}_i)} \exp(\delta_{y_j,k} s(g_k(\boldsymbol{x}_i)) g^\#(\boldsymbol{x}_i)).$$
$$(8.29)$$

Denote the right side of Eq. (8.29) as $\mathscr{A}(\boldsymbol{\Delta}|\boldsymbol{\theta})$, which is a lower bound to $T(\boldsymbol{\theta}+\boldsymbol{\Delta}) - T(\boldsymbol{\theta})$. Setting the derivative of $\mathscr{A}(\boldsymbol{\Delta}|\boldsymbol{\theta})$ w.r.t. $\delta_{y_j,k}$ to zero gives

$$\frac{\partial \mathscr{A}(\boldsymbol{\Delta}|\boldsymbol{\theta})}{\partial \delta_{y_j,k}} = \sum_i d_{\boldsymbol{x}_i}^{y_j} g_k(\boldsymbol{x}_i)$$
$$- \sum_i p(y_j|\boldsymbol{x}_i; \boldsymbol{\theta}) g_k(\boldsymbol{x}_i) \exp\left(\delta_{y_j,k} s(g_k(\boldsymbol{x}_i)) g^\#(\boldsymbol{x}_i)\right) = 0. \quad (8.30)$$

Algorithm 1: IIS-LLD

Input: The training set $S = \{(\boldsymbol{x}_i, D_i)\}_{i=1}^n$, the feature functions $g_k(\boldsymbol{x})$, and the convergence
  criterion ε
Output: $p(y|\boldsymbol{x}; \boldsymbol{\theta})$

1 Initialize the model parameter vector $\boldsymbol{\theta}^{(0)}$;
2 $i \leftarrow 0$;
3 **repeat**
4 | $i \leftarrow i + 1$;
5 | Solve Eq. (8.30) for $\delta_{y,k}$;
6 | $\boldsymbol{\theta}^{(i)} \leftarrow \boldsymbol{\theta}^{(i-1)} + \boldsymbol{\Delta}$;
7 **until** $T(\boldsymbol{\theta}^{(i)}) - T(\boldsymbol{\theta}^{(i-1)}) < \varepsilon$;
8 $p(y|\boldsymbol{x}; \boldsymbol{\theta}) \leftarrow \frac{1}{Z} \exp\left(\sum_k \theta_{y,k}^{(i)} g_k(\boldsymbol{x})\right)$;

What is nice about Eq. (8.30) is that $\delta_{y,k}$ appears alone, and therefore can be solved one by one through nonlinear equation solvers, such as the Gauss-Newton method. This algorithm is named as IIS-LLD (i.e., IIS—Learning from Label Distributions) and summarized in Algorithm 1.

After $p(y|x)$ is learned from the training set, the label distribution of any new instance x' can be generated by $p(y|x')$. The availability of the explicit label distribution for x' provides many possibilities in classifier design. To name just a few, if the expected class label for x' is single, then the predicted label could be $y^* = \mathrm{argmax}_y\, p(y|x')$, together with a confidence measure $p(y^*|x')$. If multiple labels are allowed, then the predicted label set could be $L = \{y | p(y|x') > \xi\}$, where ξ is a predefined threshold. Moreover, all the labels in L can be ranked according to their description degrees. For the problem of exact age estimation, the predicted age could be the one with the maximum description degree. For the problem of age range estimation, the predicted age range could be the one with the maximum sum of description degrees of all the ages within an age range.

8.4 Experiments

8.4.1 Methodology

The FG-NET Aging Database [21] and the MORPH database [27] are used in the experiments. The FG-NET Aging Database contains 1,002 face images from 82 subjects. Each subject has 6–18 face images at different ages. Each image is labeled by its chronological age. The ages are distributed in a wide range from 0 to 69. Besides age variation, most of the age-progressive image sequences display other types of facial variations, such as significant changes in pose, illumination, expression, *etc.* A typical aging face sequence in this database is shown in Fig. 8.1. In the MORPH database, there are 55,132 face images from more than 13,000 subjects. The average number of images per subject is four. The ages of the face images range from 16 to 77 with a median age of 33. The faces are from different races, among which the African faces account for about 77 %, the European faces account for about 19 %, and the remaining 4 % includes Hispanic, Asian, Indian, and other races. Some typical aging faces in this database are shown in Fig. 8.7.

There are different ways to extract features from the face images. For each database, we select the feature extractor from several options that can achieve the best performance. The feature extractor used for the FG-NET database is the Appearance Model [6]. The main advantage of this model is that the extracted features combine the shape and intensity of the face images, both of which are important in the aging progress. In this experiment, the first 200 model parameters are used as the extracted features. The features used for the MORPH database are the Biologically Inspired Features (BIF) [17]. By simulating the primate visual system, BIF has shown good performance in facial age estimation [17]. The dimensionality of the BIF vectors is further reduced to 200 using Marginal Fisher Analysis (MFA) [35].

As typical examples of the age estimation algorithms based on special data representation forms, AGES and IIS-LLD are compared with several existing algorithms that assume the basic data representation $x \rightarrow y$. Among the baseline methods, some

Fig. 8.7 Typical aging faces of two subjects in the MORPH database [27]. The chronological ages are given under the images

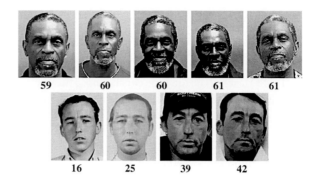

are specially designed for the age estimation problem, i.e., WAS [21], and AAS [20]. Some are conventional general-purpose classification methods, i.e., kNN (k-Nearest Neighbors), BP (Backpropagation neural network), C4.5 (C4.5 decision tree), SVM (Support Vector Machine), and a fuzzy classifier ANFIS (Adaptive-Network-Based Fuzzy Inference System) [18]. For all of these general-purpose methods, age estimation is formulated as a standard multi-class classification problem.

In AGES, the dimensionality of the aging pattern subspace is set to 20 ($d = 20$), the maximum iteration τ is set to 50, and the error threshold $\varepsilon = 10^{-3}$. As to IIS-LLD in Algorithm 1, the parameter vector θ is initialized as a zero vector, and the error threshold ε is set to 10^{-4}. According to the chronological age of each face image, a label distribution is generated using the Gaussian or triangle distribution shown in Fig. 8.6. The predicted age for a test image x' is determined by $y^* = \text{argmax}_y \, p(y|x')$. To study the usefulness of the adjacent ages, IIS-LLD is also applied to the special label distribution of the single-label case shown in Fig. 8.5a. The three kinds of label distributions are denoted by "Gaussian," "Triangle," and "Single," respectively. When generating the label distributions, the standard deviation of the "Gaussian" distribution varies within four different values 1, 2, 3, and 4. The bottom length of the "Triangle" distribution also varies within four different values 4, 6, 8, and 10. All of these label distribution settings are tested and the best results are reported.

For all the baseline algorithms, several parameter configurations are tested and the best results are reported. In AAS, the error threshold in the appearance cluster training step is set to 3. For kNN, k is set to 30 and Euclidean distance is used to find the neighbors. The BP neural network has a hidden layer of 100 neurons with sigmoid activation functions. The parameters of C4.5 are set to the default values of the J4.8 implementation (i.e., the confidence threshold 0.25 for pruning and minimum two instances per leaf). SVM is implemented as the "C-SVC" type in LIBSVM using the RBF kernel with the inverse width of 1. Finally, the number of membership functions in ANFIS is set to 2.

The performance of the age estimators is evaluated by MAE (Mean Absolute Error), i.e., the average absolute difference between the estimated age and the chronological age. The algorithms are tested through the LOPO (Leave-One-Person-Out) mode [14] on the FG-NET database, i.e., in each fold, the images of one person are used as the test set and those of the others are used as the training set. After 82 folds, each subject has been used as test set once, and the final results are calculated from all the estimates. Since there are more than 13,000 subjects in the MORPH database, the LOPO test will be too time consuming. Thus the algorithms are tested through the 10-fold cross validation on the MORPH database, i.e., the dataset is first randomly divided into 10 subsets, then in each fold, one subset is used as the test set and the rest are used as the training set. The final result is the average performance over the 10 folds.

As an important baseline, the human ability in age perception is also tested. About 5 % of the images from the FG-NET database (i.e., 51 face images) and 60 images from the MORPH database are uniformly sampled from the age ranges shown in Table 8.4. These images are used as the test samples presented to the human testees. All the testees are Chinese students or staff members from the author's university. Some other ground truth of the human tests, including the number of test samples, the number of testees, and the testees' own age, is shown in Table 8.2.

There are two stages in the human tests. In each stage, the images are randomly presented to the testees, and the testees are asked to choose one age from a given range (0–69 for FG-NET and 16–77 for MORPH) for each image. The difference between the two stages is that in the first stage (HumanA), only the gray-scale face regions (i.e., the color images are converted to the gray-scale images and the background scene of the images is removed) are shown, while in the second stage (HumanB), the whole color images are shown. Figure 8.8 gives an example of the same face shown in the HumanA test and HumanB test, respectively. HumanA intends to test the age estimation ability purely based on the intensity of the face image, which is also the input to the algorithms, while HumanB intends to test the age estimation ability based on multiple traits including face, hair, skin color, clothes, background scene, etc.

In both the HumanA and HumanB tests, each testee is required to label all the test samples, and the MAE of each testee is recorded. The minimum, maximum, and average MAE of all the testees involved in each test are given in Table 8.2. The average MAE can be regarded as a measurement of the human accuracy in age estimation. As can be seen, the testees perform remarkably better in the HumanB test than in the HumanA test, which indicates that the additional information (hair, skin color, clothes, background, etc.) provided in the HumanB test is helpful to improve the human accuracy in age estimation.

8.4.2 Results

The MAEs of all the age estimators are tabulated in Table 8.3. The standard deviations on the MORPH database are also given in the table. Note that the number of images for each person in the FG-NET database varies dramatically. Consequently, the standard deviation of the LOPO test on the FG-NET database becomes unstable. So it is

Table 8.2 Human tests on age perception

Dataset	# Samples	# Testees			Testees' age			MAE (HumanA)			MAE (HumanB)		
		Males	Females	Total	Minimum	Maximum	Average	Minimum	Maximum	Average	Minimum	Maximum	Average
FG-NET	51	24	5	29	22	44	25	4.88	15.67	8.13	4.14	8.33	6.23
MORPH	60	28	12	40	16	64	26	5.47	13.28	8.24	4.78	11.03	7.23

Fig. 8.8 An example of
the same face shown in **a**
the HumanA test, and **b** the
HumanB test

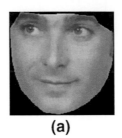

(a)   **(b)**

Table 8.3 MAE of different age estimators

Method		Dataset	
		FG-NET	MORPH
IIS-LLD	Gaussian	**<u>5.77</u>** (1, 1)	**<u>5.67 ± 0.15</u>** (1, 1)
	Triangle	**<u>5.90</u>** (1, 0)	**<u>6.09 ± 0.14</u>** (1, 1)
	Single	**6.27** (1, 0)	**<u>6.35 ± 0.17</u>** (1, 1)
AGES		**6.77** (1, 1)	**<u>6.61 ± 0.11</u>** (1,1)
WAS		**8.06** (0, 1)	9.21 ± 0.16 (1, 1)
AAS		14.83 (1, 1)	10.10 ± 0.26 (1, 1)
*k*NN		8.24 (0, 1)	9.64 ± 0.24 (1, 1)
BP		11.85 (1, 1)	12.59 ± 1.38 (1, 1)
C4.5		9.34 (1, 1)	**7.48 ± 0.12** (1, 0)
SVM		**7.25** (1, 1)	**7.34 ± 0.17** (1, 0)
ANFIS		8.86 (0, 1)	9.24 ± 0.17 (1, 1)
Human	HumanA	8.13	8.24
Tests[a]	HumanB	6.23	7.23

The MAEs of the algorithms higher than that of HumanA are highlighted by boldface and those
higher than that of HumanB are underlined[a] The human tests are performed on 5 % samples from
the FG-NET database and 60 samples from the MORPH database

not shown in Table 8.3. The MAEs of the algorithms higher than that of HumanA
are highlighted by boldface and those higher than that of HumanB are underlined.
Since the results of the human tests are the mean MAEs of multiple testees, the
two-tailed *t*-tests at the 5 % significance level are performed to see whether the
differences between the results of the human tests and the algorithms are statistically
significant. The results of the *t*-tests are given in the brackets right after the MAE of
each algorithm in Table 8.3. The number "1" represents significant difference, "0"
represents otherwise. The first number is the *t*-test result on HumanA, the second is
that on HumanB.

As can be seen, although the algorithms perform differently on the two datasets due to different test protocols and source databases, the performance of AGES and IIS-LLD is significantly better than that of the baseline methods. On the one hand, the advantage of AGES mainly comes from the data structure of aging pattern defined in Sect. 8.2.1, which naturally integrates both personal identity and age. The data representation of AGES well matches the characteristics of facial aging effects mentioned in Sect. 8.1. On the other hand, the good performance of IIS-LLD is mainly due to two reasons. First, the prior label distributions of the training examples make it possible that one instance contributes to the learning of multiple classes. Second, as discussed in Sect. 8.3.2.1, the label distribution learning algorithms tend to learn the similarity among the neighboring ages, no matter what the (reasonable) prior label distribution might be. The second reason also explains why the "Single" case of IIS-LLD can achieve state-of-the-art results even when the prior label distribution in this case is equivalent to single label. Refer back to Eq. (8.18), the learning target of the "Single" case is to ensure the dominating position of the chronological age in the label distribution. Although no prior knowledge about the neighboring ages is given, the label distribution learning algorithms can learn it based on the similarity of the face images at the close ages.

In all cases, AGES and IIS-LLD perform significantly better than HumanA. The "Gaussian" case of IIS-LLD on FG-NET and all cases of IIS-LLD and AGES on MORPH perform even significantly better than HumanB. Considering that more information is actually provided to the human testees in the HumanB test, it can be concluded that *under the experimental settings of this chapter*, IIS-LLD and AGES can both achieve better performance than that of the human testees. However, it would be too optimistic to claim that the algorithms can outperform humans in general. The main reason is that people usually perform better for faces belonging to their own race than for those belonging to another race [4]. While most images in the FG-NET and MORPH databases are Caucasian and African faces, the testees involved in the human tests are all Chinese. Thus the results of the human tests are actually biased toward a more difficult task: estimate the age of the faces from a different race.

AGES relies on the special data structure aging pattern, which is composed by all the aging faces of one person. The quantity of the training examples determines the completeness of the aging patterns, and thus affects the performance of AGES. IIS-LLD, however, is motivated by the problem of insufficient training data. By utilizing the face images at the neighboring ages when learning a particular age, IIS-LLD can effectively relieve this problem. To reveal the effect of the quantity of the training examples, IIS-LLD is compared with AGES in different age ranges on the FG-NET database. The results are tabulated in Table 8.4. As can be seen, the number of examples in different age ranges decreases rapidly with the increase of age. The examples in the higher age groups (e.g., 60–69) are especially rare. It is interesting to find that in the age ranges with relatively sufficient training data, the performance of IIS-LLD could be worse than that of AGES. For example, in the age ranges 0–9 and 10–19, all the three cases of IIS-LLD perform worse than AGES. This is because that IIS-LLD is based on the general-purpose maximum entropy model, while AGES builds on the problem-specific data structure aging pattern.

Table 8.4 MAE in different age ranges on the FG-NET database

Range	# Examples	IIS-LLD			AGES
		Gaussian	Triangle	Single	
0–9	371	2.83	2.83	3.06	2.30
10–19	339	5.21	5.17	4.99	3.83
20–29	144	6.60	6.39	6.72	8.01
30–39	79	11.62	11.66	12.10	17.91
40–49	46	12.57	15.78	18.89	25.26
50–59	15	21.73	22.27	27.40	36.40
60–69	8	24.00	26.25	32.13	45.63

The advantage of the problem-specific model generally becomes more apparent when there are sufficient training data. Another important fact revealed by Table 8.4 is that the main advantage of IIS-LLD comes from the classes with insufficient training examples. The less training examples there are, the more apparent the superiority of IIS-LLD becomes. For example, in the age range 0–9 with maximum number of training examples, the MAE of the "Gaussian" case of IIS-LLD is 23 % higher than that of AGES, while in the age range 60–69 with minimum number of training examples, the MAE of the "Gaussian" case of IIS-LLD becomes 47 % lower than that of AGES.

8.5 Summary

The aging of human faces is an interesting process with several characteristics, such as personal, temporal, gradual, and uncontrollable. Although most existing methods for facial age estimation still follow the traditional way to model the $x \rightarrow y$ mapping, the characteristics of aging progress might require matching data representations to get better performance. This chapter tries to solve the facial age estimation problem from the data representation perspective. Two typical methods based on data representations specially designed according to the characteristics of the aging progress are presented in this chapter. The first is AGES, which is based on a special data structure called aging pattern. An aging pattern is a sequence of personal face images sorted in time order, which well matches the personal and temporal characteristics. The second is based on a label distribution learning algorithm IIS-LLD. A label distribution covers not only the chronological age, but also the neighboring ages, which makes it possible to utilize the face images at the neighboring ages while learning a particular age. This well matches the gradual characteristic and relieves the insufficient training example problem caused by the uncontrollable characteristic. Experimental results on two aging face databases (FG-NET and MORPH) show clear advantages of AGES and IIS-LLD over several existing age estimation methods.

Acknowledgments This work was supported by the National Science Foundation of China (61273300, 61232007), the Scientific Research Foundation for the Returned Overseas Chinese Scholars, State Education Ministry, the Excellent Young Teachers Program of SEU, and the Key Lab of Computer Network and Information Integration of Ministry of Education of China.

References

1. Berger, A.L., Pietra, S.D., Pietra, V.J.D.: A maximum entropy approach to natural language processing. Computat. Linguist. **22**(1), 39–71 (1996)
2. Bruyer, B., Scailquin, J.C.: Person recognition and ageing: the cognitive status of addresses—an empirical question. Int. J. Psychol. **29**(3), 351–366 (1994)
3. Chang, K.Y., Chen, C.S., Hung, Y.P.: Ordinal hyperplanes ranker with cost sensitivities for age estimation. In: Proceedings of IEEE Conference on Computer Vision and Pattern Recognition, pp. 585–592. Colorado Springs (2011)
4. Dehon, H., Brédart, S.: An 'other-race' effect in age estimation from faces. Perception **30**(9), 1107–1113 (2001)
5. Denoeux, T., Zouhal, L.M.: Handling possibilistic labels in pattern classification using evidential reasoning. Fuzzy Sets Syst. **122**(3), 409–424 (2001)
6. Edwards, G.J., Lanitis, A., Cootes, C.J.: Statistical face models: improving specificity. Image Vision Comput. **16**(3), 203–211 (1998)
7. FG-NET Aging Database. http://sting.cycollege.ac.cy/~alanitis/fgnetaging/index.htm
8. Fu, Y., Guo, G., Huang, T.S.: Age synthesis and estimation via faces: a survey. IEEE Trans. Pattern Anal. Mach. Intell. **32**(11), 1955–1976 (2010)
9. Fu, Y., Huang, T.: Human age estimation with regression on discriminative aging manifold. IEEE Trans. Multimedia **10**(4), 578–584 (2008)
10. Fu, Y., Xu, Y., Huang, T.S.: Estimating human age by manifold analysis of face pictures and regression on aging features. In: Proceedings of IEEE International Conference on Multimedia and Expo, pp. 1383–1386. Beijing (2007)
11. Geng, X., Smith-Miles, K., Zhou, Z.H.: Facial age estimation by learning from label distributions. In: Proceedings of 24th AAAI Conference on Artificial Intelligence, pp. 451–456. Atlanta (2010)
12. Geng, X., Yin, C., Zhou, Z.H.: Facial age estimation by learning from label distributions. IEEE Trans. Pattern Anal. Mach. Intell. **35**(10), 2401–2412 (2013)
13. Geng, X., Zhou, Z.H., Smith-Miles, K.: Automatic age estimation based on facial aging patterns. IEEE Trans. Pattern Anal. Mach. Intell. **29**(12), 2234–2240 (2007)
14. Geng, X., Zhou, Z.H., Zhang, Y., Li, G., Dai, H.: Learning from facial aging patterns for automatic age estimation. In: Proceedings of the 14th ACM International Conference on Multimedia, pp. 307–316. Santa Barbara (2006)
15. Guo, G., Fu, Y., Dyer, C.R., Huang, T.S.: Image-based human age estimation by manifold learning and locally adjusted robust regression. IEEE Trans. Image Process. **17**(7), 1178–1188 (2008)
16. Guo, G., Mu, G.: Simultaneous dimensionality reduction and human age estimation via kernel partial least squares regression. In: Proceedings of IEEE Conference on Computer Vision and Pattern Recognition, pp. 657–664. Colorado Springs (2011)
17. Guo, G., Mu, G., Fu, Y., Huang, T.S.: Human age estimation using bio-inspired features. In: Proceedings of IEEE Conference on Computer Vision and Pattern Recognition, pp. 112–119. Miami (2009)
18. Jang, J.S.R.: ANFIS: Adaptive-network-based fuzzy inference system. IEEE Trans. Syst. Man Cybern. B **23**(3), 665–685 (1993)
19. Jolliffe, I.T.: Principal Component Analysis, 2nd edn. Springer, New York (2002)

20. Lanitis, A., Draganova, C., Christodoulou, C.: Comparing different classifiers for automatic age estimation. IEEE Trans. Syst. Man Cybern. Part B **34**(1), 621–628 (2004)
21. Lanitis, A., Taylor, C.J., Cootes, T.: Toward automatic simulation of aging effects on face images. IEEE Trans. Pattern Anal. Mach. Intell. **24**(4), 442–455 (2002)
22. Leonardis, A., Bishof, H.: Robust recognition using eigenimages. Comput. Vis. Image Und. **78**(1), 99–118 (2000)
23. Ni, B., Song, Z., Yan, S.: Web image mining towards universal age estimator. In: Proceedings of the 17th ACM International Conference on Multimedia, pp. 85–94. Vancouver (2009)
24. Ni, B., Song, Z., Yan, S.: Web image and video mining towards universal and robust age estimator. IEEE Trans. Multimedia **13**(6), 1217–1229 (2011)
25. Pietra, S.D., Pietra, V.J.D., Lafferty, J.D.: Inducing features of random fields. IEEE Trans. Pattern Anal. Mach. Intell. **19**(4), 380–393 (1997)
26. Quost, B., Denoeux, T.: Learning from data with uncertain labels by boosting credal classifiers. In: Proceedings of 1st ACM SIGKDD Workshop on Knowledge Discovery from Uncertain Data, pp. 38–47. Paris (2009)
27. Ricanek, K., Tesafaye, T.: MORPH: a longitudinal image database of normal adult age-progression. In: Proceedings of 7th International Conference on Automatic Face and Gesture Recognition, pp. 341–345. Southampton (2006)
28. Roweis, S.: EM algorithms for PCA and SPCA. In: Jordan, M.I., Kearns, M.J., Solla, S.A. (eds.) Advances in Neural Information Processing Systems 10, pp. 626–632. MIT Press, Cambridge (1998)
29. Smyth, P.: Learning with probabilistic supervision. In: Petsche, T. (ed.) Computational Learning Theory and Natural Learning System, vol. III, pp. 163–182. MIT Press, MA (1995)
30. Tipping, M.E., Bishop, C.M.: Probabilistic principal component analysis. J. Roy. Stat. Soc. B: Stat. Methodol. **61**, 611–622 (1999)
31. Tsoumakas, G., Katakis, I.: Multi-label classification: an overview. Int. J. Data Warehouse. Min. **3**(3), 1–13 (2007)
32. Wiberg, T.: Computation of principal component when data are missing. In: Proceedings of the 2nd Symposium on Computational Statistics, pp. 229–236. Berlin (1976)
33. Yan, S., Wang, H., Huang, T.S., Yang, Q., Tang, X.: Ranking with uncertain labels. In: Proceedings of IEEE International Conference on Multimedia and Expo, pp. 96–99. Beijing (2007)
34. Yan, S., Wang, H., Tang, X., Huang, T.S.: Learning auto-structured regressor from uncertain nonnegative labels. In: Proceedings of IEEE International Conference on Computer Vision, pp. 1–8. Rio de Janeiro (2007)
35. Yan, S., Xu, D., Zhang, B., Zhang, H., Yang, Q., Lin, S.: Graph embedding and extensions: a general framework for dimensionality reduction. IEEE Trans. Pattern Anal. Mach. Intell. **29**(1), 40–51 (2007)
36. Yan, S., Zhou, X., Liu, M., Hasegawa-Johnson, M., Huang, T.S.: Regression from patch-kernel. In: Proceedings of IEEE Conference on Computer Vision and Pattern Recognition. Anchorage (2008)
37. Zhang, Y., Yeung, D.Y.: Multi-task warped gaussian process for personalized age estimation. In: Proceedings of IEEE Conference on Computer Vision and Pattern Recognition, pp. 2622–2629. San Francisco (2010)
38. Zhuang, X., Zhou, X., Hasegawa-Johnson, M., Huang, T.S.: Face age estimation using patch-based hidden markov model supervectors. In: Proceedings of International Conference on Pattern Recognition, pp. 1–4. Tampa (2008)
39. Zimmermann, H.J. (ed.): Practical Applications of Fuzzy Technologies. Kluwer Academic Publishers, Netherlands (1999)

Chapter 9
Identity and Kinship Relations in Group Pictures

Ming Shao, Siyu Xia and Yun Fu

9.1 Introduction

This chapter studies the usage of knowledge about kinship to make more accurate predictions about identity in group pictures. Group pictures are an important case for people identification in many applications, for example, photo albums in social media applications [7]. Group pictures present distinct challenges but also provide a lot of external information which can be leveraged [21].

Identifying people in group pictures is an important problem that has recently received increased attention. The problem of face recognition in group pictures also falls at the core of understanding media in the wild, and has received a great deal of attention in the past few years.

The proposed models in this chapter formulate the face recognition problem (given an image of a person, determine the image in a fixed gallery that corresponds to this person) as an MRF labeling problem. There are two types of edges (dependencies): *intraimage edges*, which are based on kinship similarity, and *interimage edges* that are based on identity similarity. An example of our MRF-based group photo labeling approach can be seen in Fig. 9.1.

As part of this work, a large dataset of (mostly family) group pictures in the wild has been collected, which is named as Family Groups in the Wild (FGW). It contains

M. Shao (✉)
College of Computer and Information Science, Northeastern University,
360 Huntington Ave., Boston, MA 02115, USA
e-mail: mingshao@ccs.neu.edu

S. Xia
School of Automation, Southeast University, Nanjing 210096, China
e-mail: xia081@gmail.com

Y. Fu
Department of Electrical and Computer Engineering, Northeastern University,
360 Huntington Ave., Boston, MA 02115, USA
e-mail: yunfu@ece.neu.edu

Y. Fu (ed.), *Human-Centered Social Media Analytics*,
DOI: 10.1007/978-3-319-05491-9_9, © Springer International Publishing Switzerland 2014

Fig. 9.1 Example setup for face recognition in the wild using FGW. This example has six sites ($s_1, \ldots, s_6$) and 14 labels ($l_1, \ldots, l_{14}$). The probe images are converted into an MRF, where the sites are the detected faces. The edges are relationship between the sites, either intraimage kinship-based affinity (*blue solid lines*) or interimage identity-based affinity (*red dashed lines*)

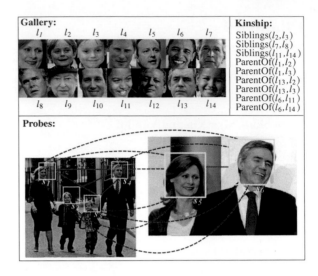

images of public personalities and their families and their kinship graph. We envision many uses for this dataset and employ it in our group images-based face identification experiments.

The task of face recognition in the wild is complicated by simultaneous variation of *pose*, *illumination*, *expression* and *age*, both in the gallery and in the probe images. Our method for face recognition builds on the observation that group pictures provide a lot of additional information, and in many practical applications, related people appear both in the images and in the gallery. We have therefore developed sensible priors about correct labeling of face images.

This chapter discusses one very interesting problem about using kinship information to increase the performance of face recognition in unconstrained settings. This is commonly referred to as face recognition in the wild [8].

The rest of the chapter is organized as follows: in Sect. 9.2 we present the previous work; our face identity and kinship models are presented in Sect. 9.3 and our problem formulation in Sect. 9.4; in Sect. 9.5 we describe our new dataset; we present our experimental evaluation in Sect. 9.6 and conclude in Sect. 9.7.

9.2 Previous Work

A great deal of work has been done in the past decades on face recognition, see [24] for an excellent survey.

This chapter builds on three streams of work. The first is on determining kinship relations from images of faces by Fang et al. [6]. This turns out to be an extremely difficult problem. As part of their work they collected a database with 143 pairs of images that are related, which we call the Cornell Kinship Verification Dataset. More recently, Shao et al. [12, 22, 23] have studied the same problem. Shao et al. provided an excellent database called UB Kinface Dataset that is designed for the study of vertical kinship relations. We build on their work to be able to handle both vertical (e.g., father–son) and horizontal (e.g., brother–sister) kinship relations. Notably, a most recent work by Fang et al. [5] that finds the appropriate family for an individual test. While they utilize kinship information (familial feature) as well, the problem discussed there is different from ours.

The second line of work is on recognizing faces obtained in challenging, unconstrained conditions. In the past few years, there has been great interest in face recognition in unconstrained settings. Consequently, researchers have produced new datasets of images acquired in unconstrained environments. One notable example of such a dataset is Labeled Faces in the Wild (LFW [8]). This is a huge collection of those images from the news, in which the Viola and Jones [20] face detector is able to find faces. Another is PubFig [10], which is collection of images of public personalities. These databases are extensive, real, difficult and have a specific evaluation protocol. We envision the database we are releasing as part of this work, called FGW, to have the same features but focused on the problem of kinship similarity.

While unconstrained face recognition has received a lot of attention, other works, that systematically study the problem of face recognition under one or a couple of variations, have been presented. Ramanathan and Chellappa [17] studied the problem of matching face images taken years apart, and proposed an adaptation of the probabilistic eigenspace framework [14]. Ling et al. [11] also studied this problem and proposed an algorithm based on learning facial differences that are described using a Gradient Orientation Pyramid (GOP).

There are also a wide variety of datasets that provide systematic variation of one or several confounding factors in face recognition. One such database obtained in an unconstrained setting is the BioID dataset [9]. It aims to capture significant variability in pose, lighting and expression. Images are captured in a realistic setting, for example, in a home environment. There are also many datasets that provide systematic variation of confounding factors obtained in controlled conditions. One of the most widely used is CMU-PIE [18], which provides systematic variation over pose, illumination and expression for 68 individuals. The Face Recognition Grand Challenge (FRGC) [16] presents a six experiment challenge problem along with a dataset of 50,000 images. The images in the dataset are collected both in controlled and uncontrolled settings.

The third is the usage of group/social information in pictures in face recognition [7, 19, 21]. Great examples of this are work by Wang et al. [21], which is similar to ours in the sense that it uses features to capture information about kinship in images. However, it does not include the direct comparison of images to determine kinship. In that respect, the methods are orthogonal and the approaches can be seen as complimentary. Social context is employed in [19] to analyze large-scale face databases captured from the Internet, e.g., Facebook, and it scales well as number of images increases. However, it only considers co-occurrence rather than recognizable relations, i.e., kin relations discussed in this chapter.

Along this same line of work, we build on the work by Manyam et al. [13], where they observe that faces from the same image often have the same illumination and similar poses and mostly similar expressions.

To summarize, while there are a wide variety of face datasets available, few of these datasets allow for the study of face recognition from group pictures of families. The collected dataset in this chapter is the first to enable the study of such cases, especially for kinships. Additionally, there have been a wide variety of algorithms proposed for face recognition in the presence of important variations such as pose, illumination, expression, and aging, but to the best of our knowledge, no one has used kinship similarity to enhance the predictions of face recognition. Our work evaluates some of these algorithms using our dataset. The availability of such a dataset from this work will encourage the development of algorithms that specifically account for family group pictures variation.

9.3 Identity and Kinship Similarity Models

In this section, our models for identity and kinship are presented with details. The key part of our models is that they have to compare a probe image to a gallery image and give a score. There are two kinds of models:

- *Identity similarity*: in this case the cost is expected to be low if comparing images of the same person.
- *Kinship similarity*: in this case the cost is expected to be low if comparing images of people in a kinship relationship. There are two types of kinship relationships: Vertical (V) and Horizontal (H).

In this chapter, we have used two methods to generate image descriptions:

- *Direction of gradient-based descriptor*: generates dense image descriptors based on the direction of gradient as described by Ling et al. [11].
- *Dense stereo matching costs-based descriptor*: generates dense image descriptors base on stereo matching costs as described by Castillo and Jacobs [3].

Process of building the descriptions will be described in the following sections.

9.3.1 Direction of Gradient

In this section, the Gradient Orientation (GO) method proposed by Ling et al. [11] is adopted to compute kinship and identity similarity. This method was designed for face recognition with aging. It uses the direction of the image gradient as an image description. This is a widely used representation that is robust to illumination changes [4]. These are compared using kernel-based learning techniques similar to the method of Phillips [15]. The method integrates Support Vector Machines (SVM) and GO.

Specifically, the SVM-GO method uses the direction of gradient as the image representation:

$$g(I(p; \kappa)) = \nabla(I(p; \kappa))/|\nabla(I(p; \kappa))|. \tag{9.1}$$

Finally, an SVM using an RBF kernel is trained using the above mentioned descriptor. As in the SVM difference method, the γ parameter of the kernel and the C parameter of the SVM are set by sampling on a grid and evaluating the performance using 5-fold cross validation on the training set.

The authors also presented a more sophisticated multiscale method called GOP. The method works slightly better (about 1% EER) than the basic SVM-GO. The GOP uses a pyramid as follows:

$$I(p; 0) = I(p), \tag{9.2}$$

$$I(p; \kappa) = [I(p; \kappa - 1) * \Phi(p)] \downarrow_2, \quad \kappa = 1 \ldots s, \tag{9.3}$$

where $\Phi(p)$ is a Gaussian kernel (0.5 is used as standard deviation), $\downarrow_2$ denotes half size down sampling and s is the number of layers in the pyramid. In initial experiments, it is found that SVM-GOP and SVM-GO have very similar performance on our dataset, so we omit results for SVM-GOP.

9.3.2 Stereo Matching

In this section, stereo matching is used to determine identity and kinship similarity, which is built on the work of Castillo and Jacobs [2] to compute a distance between images. Using this face comparison algorithm, we extract descriptors that can be used for classification. Using these models allows it to be robust to variations in pose and illumination. Figure 9.2 shows a description of the approach.

The stereo algorithm uses Dynamic Programming (DP) to find the minimum cost matching (or correspondences) between two scan lines. The images are assumed to have been rectified according to the epipolar geometry. Each step in the solution accounts for a single pixel in one of the two images. This is done using four planes (or cost matrices) called C_{Lo}, C_{Lm}, C_{Ro} and C_{Rm}. In this setup, each point in one of the matrices represents the last point in each image that has been accounted for, along with what the last step was. Points are accounted for by matching (m) and occlusions (o) in the left (L) and right (R) images.

The elements of the cost matrix are initialized to $+\infty$ at every location except:

$$C_{Ro}[i, 0] = i\alpha, \quad \forall i = 0 \ldots W - 1, \tag{9.4}$$

where α is the cost of staying in the occluded state.

The first step fills four matrices according to the following equation, in which β is the cost of beginning an occlusion, and β' is the cost of ending one:

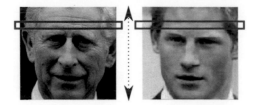

Fig. 9.2 Example of usage of stereo matching to compute the similarity of two images. The correspondences are computed one scan line at a time, and with these results a descriptor is built. The descriptor encodes pose-invariant dense similarities between both images

$$C_{Lo}[l, r] = \min(C_{Lo}[l, r - 1]$$
$$+ \alpha, C_{Lm}[l, r - 1] + \beta, C_{Rm}[l, r - 1] + \beta), \qquad (9.5)$$

$$C_{Lm}[l, r] = M(l, r) + \min \begin{cases} C_{Lo}[l, r - 1] + \beta' \\ C_{Lm}[l, r - 1] + \gamma \\ C_{Rm}[l, r - 1] \\ C_{Ro}[l, r - 1] + \beta' \end{cases}, \qquad (9.6)$$

where $M(l, r)$ is the matching cost of the lth pixel in the left scan line with the rth pixel in the right scan line. α, β, β' and γ are parameters that can be set empirically. α represents the cost of staying in the same occluded state, β represents the penalty for starting a run of occlusions. γ guards against transitions within the same matched state. The cost of matching the two scan lines l_1 and l_2, is $C_{Ro}[l - 1, r - 1]$. This cost is used to evaluate the similarity of two images. The correspondences are obtained by following a backward step. Therefore, a cost can be associated with each pixel in each image being compared. These dense costs are the descriptor D. A benefit [3] of obtaining the correspondences is we can build a descriptor of image similarities that is robust to pose and illumination change, and the method is relatively fast and practical.

9.3.3 Identity and Kinship Similarity Models/Classification

At training time, once the descriptors have been generated, a SVM classifier is trained on the same/not same task, similar to [14, 15]. At test time, when a new pair of images needs to be compared, we compute the descriptor and apply the SVM by using its signed distance to the margin as a measure of the similarity between the two images.

In this chapter, we use the descriptions from the two methods (direction of gradient and stereo matching) and data from existing datasets to train two types of models: identity similarity (denoted I^S) and kinship similarity (denoted K^S). In the following sections I^S (resp. K^S) will be expected to be positive when queried on two images of the same person (resp. kin pair), and negative otherwise.

9.4 Recognition Formulation

This section discusses how to cast the face recognition problem as an MRF MAP estimation problem (finding a labeling). We formulate a binary cost that uses kinship similarity, allowing us to optimize the complete cost using Iterated Conditional Modes (ICM) [1].

The cost to minimize is therefore:

$$E(f) = \sum_{p \in S} U(p, f_p) + \lambda \sum_{\{p,q\} \in N} d_{p,q}(f_p, f_q) + \gamma \sum_{l \in \mathbb{L}} \delta_l(f), \qquad (9.7)$$

where S is the set of sites, p and q are sites, N is the set of neighboring sites and f is a labeling, in which f_r is the label assigned to site $r \in S$ under the labeling f. U is the unary cost function and d is the binary or pairwise cost function. We also define $\delta_l(f)$ as:

$$\delta_l(f) = \begin{cases} 1 & \exists p : f_p = l \\ 0 & \text{otherwise} \end{cases}. \qquad (9.8)$$

This assigns a fixed cost to every distinct label used, and therefore discourages the use of excess labels.

9.4.1 Unary Cost

Unary cost $U(p, f_p)$ is defined as $I^S(p, g(f_p))$, where $g(f_p)$ denotes the gallery image associated with label f_p, and $I^S(\cdot, \cdot)$ is the (identity) similarity between two images.

9.4.2 Binary Cost

For pairs of sites on the same image, the binary cost $d_{p,q}(f_p, f_q)$ is defined as:

$$d_{p,q}(f_p, f_q) = \exp\left(-\gamma \cdot \max(K_H^T(f_p, f_q), K_V^T(f_p, f_q)) \cdot K^S(p, q)\right), \qquad (9.9)$$

where $K_H^T(f_p, f_q)$ is 1 if f_p and f_q satisfy a horizontal kinship relationship, -1 otherwise. Similarly, for $K_V^T(f_p, f_q)$ is 1 iff f_p and f_q satisfy a vertical kinship relationship. Additionally, we define $K^S(p, q)$ as the image-based kinship similarity between sites p and q. We call these edges intraimage kinship-based affinities.

For pairs of sites on different images, the binary cost $d_{p,q}(f_p, f_q)$ is defined as:

$$d_{p,q}(f_p, f_q) = \exp\left(-\gamma \cdot Q(f_p = f_q) \cdot I^S(p, q)\right), \tag{9.10}$$

where $Q(x) = 1$ if x is true, and -1 otherwise, and $I^S(p, q)$ is the image-based identity similarity between sites p and q. We call these edges interimage identity-based affinities.

While this binary cost expresses the requirements for the desired solution, this cost is unfortunately not submodular. We analyzed some test cases, about 40 % of the entries are non-submodular, making strategies to enforce submodularity, e.g., truncation, impractical.

9.4.3 Graph Structure and Estimation Procedure

The graph structure for this problem is a complete graph, a graph where every pair of distinct vertices is connected by an edge. We have therefore a relatively difficult, unorthodox MAP estimation problem: the graph is not a grid and the binary cost is not submodular. The problem, however, has the fundamental benefit that: (1) the number of sites is relatively small, (2) the number of labels, even in realistic cases, is fairly limited.

The problem can be solved by ICM, a simple but very well understood algorithm. It works as follows:

1. Start with a labeling L.
2. At each site, evaluate every label. Greedily set the label that gives the lowest energy value.
3. Repeat the previous step until convergence.

The energy value is guaranteed not to increase and the algorithm has been shown to converge, which usually happens within a few iterations. We now focus our attention on the two formulations:

- *One Image at a Time Formulation*: the sites (detected faces) of one image form a complete graph, with only one type of edge: intraimage, kinship-based affinity.
- *Whole Collection Formulation*: the sites (detected faces) of the whole collection form a complete graph; there are two type of edges: (1) intraimage, kinship-based affinity and (2) inter-image, similarity-based affinity. Figure 9.3 shows an example of this graph.

What model should be used depends on the expected distribution of the identities. If the identities are centered around a few individuals, then the whole collection formulation seems reasonable. If the identities are of family groups and these are mostly independent, then one image at a time formulation is a reasonable choice. In any case, the method to solve this problem is very straightforward. We use ICM to solve N independent, one image at a time, subproblems. If solving the one image at a time formulation, then this is the final result. If not (which means that the whole collection formulation is being used), we then use resulting labeling to warm start ICM to solve the whole collection formulation.

Fig. 9.3 Example case: two images, four detections on one image, three detections on the other image. This is the structure of the MRF. *Solid lines* are intraimage, kinship-based affinity edges, and *dashed lines* are interimage, similarity-based affinity edges

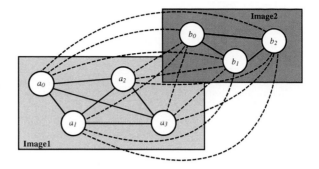

9.5 FGW Dataset

The collected dataset used in this chapter includes 460 group images mainly of families, with 105 different individuals containing a total of 1,147 faces. There is a subset of 84 individuals for which there are at least two faces. The individuals with more images are: Barack Obama (60), Rick Santorum (35), and Michelle Obama (34).

We have labeled the 1,147 images with ground truth information about their identity and their kinship. For each face, seven fiducial points: outer corner of left eye, inner corner of left eye, outer corner of right eye, inner corner of right eye, left corner of mouth, right corner or mouth and top lip are provided. We call the dataset FGW, shown in Fig. 9.4. This is the first dataset of its type, as it emphasizes both group pictures and kinship relations.

The dataset has been obtained by searching online for public personalities, such as US president Obama and former British prime minister Brown. We build the dataset around these core people and their families, as shown in Fig. 9.5.

We envision many uses for an extensive, hard, real database like FGW. In this work, we give one surprising example of usage: we show that by being able to determine kinship relations from images we can use this knowledge to increase the performance of face recognition of family pictures in unconstrained settings.

9.6 Experimental Evaluation

The two main objectives in the design of our experiments are: (1) we want to show that while relatively simple, our kinship similarity measure works surprisingly well compared to existing kinship verification methods. To do so, we evaluate on existing kinship verification datasets, and compare to existing methods as reported by the original authors; (2) we want to show that we are able to use this kinship similarity measure build a face labeling system that accounts for kinship that is significantly more accurate than using identity similarity alone.

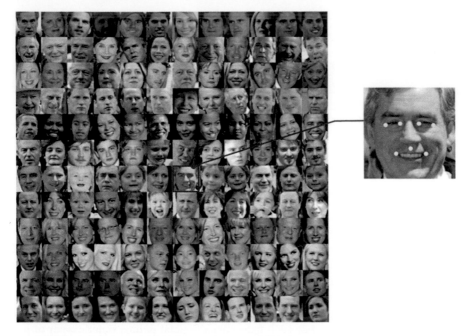

Fig. 9.4 Example faces and fiducial points from our FGW dataset

Fig. 9.5 Example of the images obtained for Barack Obama

Table 9.1 Evaluation of our kinship similarity method on the Cornell Kinship verification dataset

Method	Accuracy (%)
Left eye gray value [6]	59.7
Skin color [6]	59.7
Skin gray [6]	59.8
Left eye color [6]	60.5
Right eye color [6]	61.4
Human performance [6]	67.2
Feature integration [6]	70.5
Stereo matching [ours]	**71.3**

For objective (1) two datasets are used for experiments: the Cornell Kinship Verification Dataset [6] and the UB KinFace dataset (Xia et al. [22, 23] and Shao et al. [12]). For objective (2) we use our own FGW dataset.

In all experiments, in which we evaluate on existing datasets, our method obtains the state-of-the-art results. The details of our evaluations are presented as follows.

9.6.1 Cornell Kinship Verification Dataset Experiments

The evaluation of our kinship similarity model on the Cornell Kinship Verification Dataset [6] is organized as follows. There are 143 pairs of parent–child images. The authors also provide the coordinates of the eye locations. We use the 143 kinship pairs from the dataset and generate 143 random false kinship pairs, and evaluate the accuracy by doing fivefold cross validation.

To compute kinship similarity, we build the stereo matching descriptor and train an SVM classifier with an RBF kernel with $+1/-1$ labels based on the kinship relationship. Finally, our results are surprisingly good, to our best knowledge, achieving the state of the art on this dataset.

It is important to note that kinship verification is not the primary purpose of this work. We want to show that our kinship similarity model performs well compared to the state-of-the-art methods for kinship detection. The results using this dataset are presented in Table 9.1.

9.6.2 UB KinFace Experiments

The second evaluation for our image-based kinship similarity model is based on UB KinFace (version 2) Dataset [22, 23]. There are 200 triplets: image of child, image of young parent, image of old parent, in this dataset. The authors distribute for each image 4 feature locations: left eye, right eye, tip of nose, center of mouth.

Table 9.2 Evaluation of our kinship similarity method on the UB KinFace Dataset

Method	C versus Y (%)	C versus O (%)	C versus O (%) + Y
Raw [23]	53.9	50.6	–
Structure [23]	56.7	53.3	–
Human performance [23]	–	53.17	56.0
Face [22]	–	–	55.0
Region [22]	–	–	55.3
Face with Gabor [22]	–	–	56.1
Region with Gabor [22]	–	–	60.0
Stereo matching [ours]	**60.0**	**60.25**	**60.25**

C versus Y evaluates the accuracy when comparing the images of children with young parents. C versus O evaluates the accuracy when comparing the images of children with old parents. C versus O + Y is similar to C versus O but introduces an auxiliary dataset, i.e., young parents. Note this is especially for [22]

As in the previous experiment, we note that kinship verification is not our primary purpose, but we want to show that our kinship similarity model performs well compared to the state-of-the-art methods to determine kinship similarity.

The protocol for evaluation is the following. We perform fivefold cross validation. Namely, at each round we train with 160 positive (kinship) pairs and 160 negative (nonkinship) pairs, and similarly, at each round we test on 40 positive pairs and 40 negative pairs. Table 9.2 shows the results of our method on the UB KinFace dataset. Our results suggest that we clearly outperform the method of Xia et al. [23] with respect to precision of the kinship similarity. While we significantly outperform the results of [23], we have to mention that there is clearly a lot of room for improvement in this difficult problem.

Observe that the difference between Region with Gabor [22] and our method is very small. However, it should be pointed out that our method is simpler and we get results that are as good without the need for images of the young parents which in the long term seem unreasonable to have.

9.6.3 FGW Experiments

We now describe our experiments on the FGW dataset with three methods.

1. *Unary only (baseline)*: the baseline method uses the identity similarity used in the unary cost to assign to each face the label that obtained the lowest similarity cost.
2. *Unary + binary, one image at a time*: we use both the unary cost and binary cost and perform MRF MAP estimation.
3. *Unary + binary, whole collection at once*: we use both the unary cost and binary cost and perform MRF MAP estimation, and do so by labeling the whole collection at once.

Table 9.3 Comparison of face recognition accuracy with different methods on FGW-easy, FGW-medium and FGW-hard

Case	Method	DOG (%)	SM (%)
	Unary only	25	31.25
Easy	Unary + binary (one)	37.5	56.5
	Unary + binary (all)	37.5	62.5
	Unary only	25.85	32.20
Medium	Unary + binary (one)	30.24	41.95
	Unary + binary (all)	30.24	42.93
	Unary only	18.43	26.78
Hard	Unary + binary (one)	23.38	34.73
	Unary + binary (all)	23.92	36.12

DOG: direction of gradient, SM: stereo matching. Note that "one" means one image at a time, while all means "whole collection"

Under each of the methods, two types of unary terms are evaluated: direction of gradient and stereo under Normalized SSD features.

The evaluation protocol is as follows. First, a gallery with the individuals in the collection is built. Then, several collections of increasing difficulty are generated. For each face in the dataset we compute its identity using its associated group picture, we then report the accuracy over all faces as a percentage of the total.

We now describe the dataset settings on which we evaluate our methods:

- *FGW-easy*: we identify a subset of FGW which we call easy. It is a collection of images of the Obama family. They are a total of 12 group images, in which 32 faces appear. The gallery for this experiment has 31 people, however there are only four different individuals in the probe. This track of the experiments have the fundamental benefit that allows us to evaluate current algorithms and our new method on the simplest possible subset of FGW.
- *FGW-medium*: we identify a subset of FGW which we call medium. It is a collection of 83 group pictures containing 205 face images of gallery of 31 individuals. This is a more realistic test track. They are no longer images of one family group, but multiple family groups with images of 1–20 individuals.
- *FGW-hard*: we use all the individuals for which we have two or more images. The gallery size is 84 individuals, the probe size is 1,117 faces. This is the largest scale experiment that can be done with this dataset.

All the combinations of methods under the three dataset settings are evaluated. The results are presented in Table 9.3.

Under the FGW-easy subset, we observe the significant gains from the MRF model, and also notice the gains from using the whole collection labeling.

Under the FGW-medium subset, it is observed that as the data is not as centered on one family group, the gain from using the MRF-based models is slightly smaller, but still significant. The difference between one at a time and whole collection labeling is smaller now, meaning that for this dataset the independence between labeling of different images is a sensible assumption.

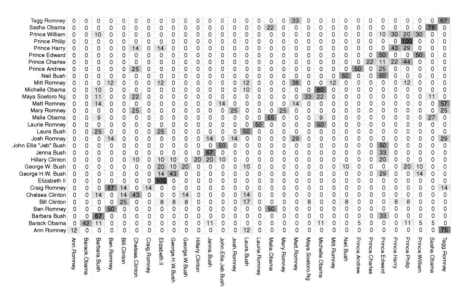

Fig. 9.6 Confusion matrix of the FGW-medium using stereo matching and whole collection prediction. The entry of ith row and jth column denotes the false alarm rate to ith individual while predicting jth individual

The confusion matrix for this test case is presented in Fig. 9.6. We can observe in the confusion matrix still some gross errors, many of which could be fixed using gender and age attributes. With gender and age attributes we could make both the identity similarity and the kinship similarity more precise.

Under the FGW-hard subset, it shows that, as expected, the problem becomes more difficult as more people are added to the gallery. Additionally, there are still gains from doing whole collection labeling. As in all other cases, there is a significant gain from unary only to any of the unary + binary models.

For each subset, we use the McNemar test to evaluate if one method is statistically significantly better than another. In all cases the gains from using the unary+binary model (instead of the unary model) are statistically significant.

9.7 Summary

In this chapter, a novel method to label faces in photo collections was proposed. The method uses kinship similarity to determine the compatibility of pairwise labels in the same image and identity similarity to determine the compatibility of faces between different images. Experiments demonstrated that our method leads to increased performance with respect to identity (recognition rate). Instead of labeling one face at a time, we show that by finding a consistent labeling we obtain significant gains in performance. Our work has the benefit that it is not connected to a particular method to compute the unary and binary terms: can be improved both by any breakthrough in determining identity similarity and kinship similarity.

Face recognition in the wild is an extremely difficult problem. While our work shows great gains in using kinship similarity as a measure of compatibility, there is obviously a lot of room for improvement. We see several important future directions for our work: usage of multisite kinship relationships, such as brother–sister–parent, father–mother–daughter, and integration with other nonimage-based features, such as age, or kinship-stable features, such as ethnicity.

References

1. Besag, J.: On the statistical analysis of dirty pictures. J. Roy. Stat. Soc. B **48**, 259–302 (1986)
2. Castillo, C.D., Jacobs, D.W.: Using stereo matching with general epipolar geometry for 2d face recognition across pose. IEEE Trans. Pattern Anal. Mach. Intell. **31**(12), 2298–2304 (2009)
3. Castillo, C.D., Jacobs, D.W.: Trainable 3d recognition using stereo matching. In: ICCV Workshop on 3D Representation and Recognition (3dRR), pp. 625–631 (2011)
4. Chen, H.F., Belhumeur, P.N., Jacobs, D.W.: In search of illumination invariants. In: IEEE Conference on Computer Vision and Pattern Recognition, pp. 1254–1261 (2000)
5. Fang, R., Gallagher, A.C., Chen, T., Loui, A.: Kinship classification by modeling facial feature heredity. In: IEEE International Conference on Image Processing. IEEE (2013)
6. Fang, R., Tang, K.D., Snavely, N., Chen, T.: Towards computational models of kinship verification. In: IEEE International Conference on Image Processing (2010)
7. Gallagher, A.C., Chen, T.: Understanding images of groups of people. In: IEEE Conference on Computer Vision and Pattern Recognition, pp. 256–263. IEEE (2009)
8. Huang, G.B., Ramesh, M., Berg, T., Learned-Miller, E.: Labeled faces in the wild: a database for studying face recognition in unconstrained environments. In: Faces in Real-Life Images Workshop in ECCV (2008)
9. Jesorsky, O., Kirchberg, K., Frischolz, R.: Robust face detection using the hausdorff distance. In: Audio and Video Based Person Authentication, pp. 90–95 (2001)
10. Kumar, N., Berg, A.C., Belhumeur, P.N., Nayar, S.K.: Attribute and simile classifiers for face verification. In: International Conference on Computer Vision, pp. 365–372 (2009)
11. Ling, H., Soatto, S., Ramanathan, N., Jacobs, D.W.: A study of face recognition as people age. In: International Conference on Computer Vision, pp. 1–8 (2007)
12. Shao, M., Xia, S., Fu, Y.: Genealogical face recognition based on ub kinface database. In: CVPR Workshop on Biometrics (BIOM) (2011)
13. Manyam, O.K., Kumar, N., Belhumeur, P.N., Kriegman, D.J.: Two faces are better than one: face recognition in group photographs. In: International Joint Conference on Biometrics (2011)
14. Moghaddam, B., Jebara, T., Pentland, A.: Bayesian modeling of facial similarity. In: Advances in Neural Information Processing Systems, pp. 910–916 (1999)
15. Phillips, P.J.: Support vector machines applied to face recognition. In: Advances in Neural Information Processing Systems 11, pp. 803–809. MIT Press (1998)
16. Phillips, P.J., Flynn, P.J., Scruggs, T., Bowyer, K.W., Worek, W.: Preliminary face recognition grand challenge results. In: IEEE International Conference on Automatic Face and Gesture Recognition, pp. 15–24 (2006)
17. Ramanathan, N., Chellappa, R.: Face verification across age progression. IEEE Trans. Image Process. **15**(11), 3349–3362 (2006)
18. Sim, T., Baker, S., Bsat, M.: The CMU pose, illumination, and expression database. IEEE Trans. Pattern Anal. Mach. Intell. **25**(12), 1615–1618 (2003)

19. Stone, Z., Zickler, T., Darrell, T.: Toward large-scale face recognition using social network context. Proc. IEEE **98**(8), 1408–1415 (2010)
20. Viola, P., Jones, M.J.: Robust real-time face detection. Int. J. Comput. Vis. **57**(2), 137–154 (2004)
21. Wang, G., Gallagher, A.C., Luo, J., Forsyth, D.A.: Seeing people in social context: recognizing people and social relationships. In: European Conference on Computer Vision, pp. 169–182 (2010)
22. Xia, S., Shao, M., Fu, Y.: Kinship verification through transfer learning. In: International Joint Conference on Artificial Intelligence, pp. 2539–2544 (2011)
23. Xia, S., Shao, M., Luo, J., Fu, Y.: Understanding kin relationships in a photo. IEEE Trans. Multimedia **14**(4), 1046–1056 (2012)
24. Zhao, W., Chellappa, R., Phillips, P.J., Rosenfeld, A.: Face recognition: a literature survey. ACM Comput. Surv. **35**(4), 399–458 (2003)

Chapter 10
Recognizing Occupations Through Probabilistic Models: A Social View

Ming Shao and Yun Fu

10.1 Introduction

This chapter is devoted to the problem of occupation recognition of human from photos. Social characteristics are keys in parsing social relations, since they can well describe, and explain human's social behavior. Due to the rapid development of multimedia in the past decade, people attempt to exploit digital carriers to record or even analyze social characteristics. This emerging research area, namely, social media analytics, attracts substantial attention from researchers. One subproblem of social media, i.e., demographic profiles analysis has long been discussed for the past few decades, e.g., face [35], gender [4], age [14]. However, research associate with demographical profiles continues to motivate huge amount of applications on social characteristics, in particular with the emergence of many social network websites, e.g., Facebook, Twitter, Google+, and photo sharing websites, e.g., Flickr, Google Picasa. To this end, researchers explore the correlations among people: detecting multiple people by spatial layout [10]; identifying people by shot time of images, fix pattern of co-occurrence, and re-occurrence [22]; recognizing a group of people via social norm and conventional positioning [15]; jointly solving social relation and people identification [31]; parsing social interaction by first-person video [12]; recognizing people by linked text in videos or images [2]; discovering social roles in assigned scenarios [25]; exploring kin relationship via transfer learning and semantic-level context [33].

Similar to above works mining the web available profiles, we envision a novel usage of existing web accessible media. A typical scenario is: social websites are

M. Shao (✉)
College of Computer and Information Science, Northeastern University, 360 Huntington Ave, Boston, MA 02115, USA
e-mail: mingshao@ccs.neu.edu

Y. Fu
Department of Electrical and Computer Engineering, Northeastern University, 360 Huntington Ave, Boston, MA 02115, USA
e-mail: yunfu@ece.neu.edu

Y. Fu (ed.), *Human-Centered Social Media Analytics*,
DOI: 10.1007/978-3-319-05491-9_10, © Springer International Publishing Switzerland 2014

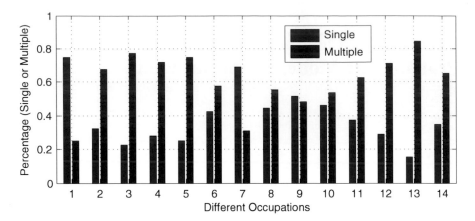

Fig. 10.1 Percentage of single and multiple person(s) images in each category: 1. chef, 2. clergy, 3. construction labor, 4. doctor, 5. firefighter, 6. lawyer, 7. mailman, 8. marathoner, 9. policeman, 10. soccer player, 11. soldier, 12. student, 13. teacher, and 14. waiter. We try to avoid bias during data collection, and download images based on the default order returned by the search engine

able to access personal profiles as well as uploaded photos, which are adequate to infer social characteristics or personal preference. It is known that people with similar background or occupations are more likely to talk and make friends. If we can recognize people's occupations based on digital carrier, more specifically, visual information in the uploaded photo, then social websites can accurately target potential customers and provide high-quality service on connection recommendation, friends impression, etc.

10.1.1 Related Work

The most related work is occupation prediction of single person in a photo where human clothing feature, foreground, and background context are exploited [29]. The preliminary results demonstrate that occupation prediction is solvable under the assumption of nearly frontal pose of people. In that work, clothing feature is described via patches from body parts, and is semantically represented by informative and noise-tolerant sparse coding [34]. In addition, both foreground and background feature is extracted by Bag-of-Words (BoWs) model [13] to complement the clothing feature, which explicitly models the social interaction. However, there are still problems unsolved that we will address in this chapter:

First, the assumption of nearly frontal pose and single person is not always strictly satisfied in real-world applications, as the ratio of single and multiple persons images for each category shown in Fig. 10.1.

Second, clothing feature is necessary but not sufficient. Some semantic-level features, e.g., age, skin color, spatial relations are also helpful on inference of the occupation category.

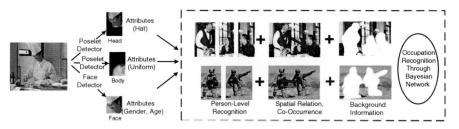

Fig. 10.2 Framework of the proposed method

Third, a person's occupation is tightly connected to others in the image, and jointly determined by both feature of itself and relation with others in the photo. Leveraging this helps to improve the performance of independent recognition.

10.1.2 Our Contributions

In this chapter, a novel framework toward multiple people occupation recognition in a more general case is proposed, as Fig. 10.2 illustrated, which is an extension of our previous work [28]. The contributions of this chapter are:

First, we use poselet [3, 5], a segmentation of human body under arbitrary pose, to capture the low-level feature, so the nearly frontal upper-body constraint is naturally removed.

Second, semantic-level features, i.e., visual attributes, instead of low-level features are adopted as the input of person-level occupation classifier. These attributes, usually human-designed meaningful tags, have been extensively studied in [11, 17, 18, 23, 30, 32]. We particularly use them to indicate semantic-level features for occupation recognition.

Third, we consider multiple people in a single photo and jointly determine their occupations at the same time via a Bayesian network. This is inspired by the observation that people of correlated occupations, e.g., teacher and student, or the same occupation, e.g., soldier and soldier, appear in the same photo with high probability. We show some statistical results based on our collected occupation data from the Internet in Fig. 10.1. In addition, the spatial correlation of different people in some scenarios can offer more cues. For example, customers are often sitting beside those who serve.

10.2 Database

The database used in this chapter includes 14 different occupations categories and over 7K images, all of which are downloaded from the Internet using image search engines, e.g., Google Image, video website, e.g., YouTube, and social network

Chef	Clergy	Construction Labor	Doctor	Firefighter
Lawyer	Mailman	Marathoner	Policeman	Soccer Player
Soldier	Student	Teacher	Waiter	

Fig. 10.3 Illustrations of our newly collected occupation database. There are 14 occupations and over 7,000 images in this database

website, e.g., Facebook. The 14 categories are selected based on the following criteria: (1) appearance feature should be visually informative; (2) images can be handled by human detectors to some extent, without dense crowd and severe overlap among people. These categories are: **chef, clergy, construction labor, doctor, firefighter, lawyer, mailman, marathoner, policeman, soccer player, soldier, student, teacher, and waiter**. We try both occupation names directly and their synonyms to obtain highly relevant photos. For the labeling work of occupation category and visual attributes therein, three professionals who have experience on image annotation are invited to label these images, based on the majority voting principle. After the image labeling as well as cleaning, we finally obtain at least 500 images for each category. The sample images of the collected database can be found in Fig. 10.3.

10.3 Data Representation via Attributes

10.3.1 Poselet-Based Appearance Modeling

Poselet body-part detectors [3, 5] and their relevant works have been successfully applied to vision problems, e.g., object segmentation [6], action recognition [21], and attributes classification [4]. Poselets offer a robust scheme to detect human

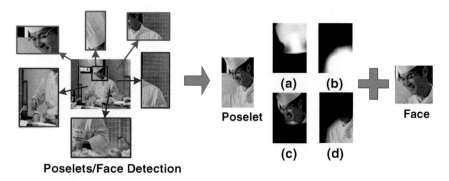

Poselet

Face

Poselets/Face Detection

Fig. 10.4 Poselets detection and masking. In the *left* figure, several poselets and a face are detected from the *center* image. The poselets in blue frames are what we defined as "head poselets," while those in *red* are "upper-body poselets." In the *right* figure, different masks are used to post-process one poselet to remove irrelevant factors. For this upper-body poselet, we use two different masks, i.e., hat mask, and upper-body mask, which are shown in *a*, and *b*. After masking operation on the input, the poselets we need are shown in *c*, and *d*

through locating partial bodies and then jointly learning the bodies or torsos by Hough voting.

In our framework, we mainly use poselets scheme for two tasks: **first**, to extract local body parts; **second**, to locate entire body region. In [29], they use four learned key points on human body to locate semantically meaningful image patches, i.e., hat, torso, left, and right shoulder. Their method works effectively under the assumption of nearly frontal upper-body, but when heads or bodies are tilted, rotated, their key points learning method may fail. Instead, we use 150 poselets learned based on the work in [3] to detect multiple body parts. According to the spatial characteristics of each attribute that we define in this chapter, we group the poselets into two sets: head poselets and upper-body poselets, as shown in Fig. 10.4. Head poselets are responsible for attributes such as "hat," while upper-body poselets for attributes such as "hat" or "uniform." Using specific group of poselets benefits us by avoiding over-fitting in the learning process. In addition, to remove noise, background, and irrelevance, we design two masks to fit hat and upper-body poselets, as shown in Fig. 10.4.

10.3.2 Attributes for Semantic-Level Features

Attributes are descriptive words designed by people to capture the visually perceptible properties of objects. These semantically meaningful mid-level features can re-organize complex relations between low-level features and high-level labels. Recently, they have been adopted for objects' property description [11], object detection [27] and recognition [32], face verification [17], scene understanding [24], clothing retrieval [20], anomaly detection [26], on-line learning [16], and one-shot learning

Table 10.1 Selections of poselet/region, mask type and feature type for each attribute

Attribute name	Poselet/region	Mask type	Feature type
Uniform (11)	U-body poselet	U-body mask	Dense grid
Hat (7)	U-body/head poselet	Hat mask	Dense grid
Skin color	Body region	N/A	Skin feature
Skin fraction	Body region	N/A	Skin feature
Age	Face region	Face mask	Raw feature
Gender	Face region	Face mask	Raw feature

Note that "U-Body" means upper-body, and "Dense Grid" means the dense grid descriptors of HOG, LBP, and CIELAB color histogram
The numbers after "Uniform" and "Hat" denote the numbers of detailed attributes in these categories

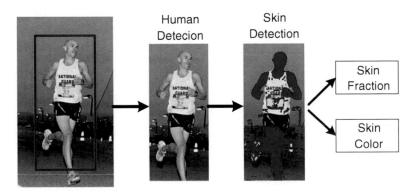

Fig. 10.5 Illustration of skin feature extraction

with wide-margin ranking function [23]. In [29], they also use mid-level features, i.e., four patches from human body, to bridge the semantic gap between features and occupation categories. However, it is far from satisfaction under complex situations such as interaction between people. Through observation, we find that not only the aforementioned attributes such as uniforms and hats play crucial roles in determination, but also other weakly relevant attributes help to jointly determine the occupation, e.g., skin features, age, and gender.

In this chapter, six attributes are proposed to represent semantic-level features that can determine or partially determine the occupations of people in a photo. These attributes are listed in Table 10.1. We categorize them into two classes: (1) strong attributes and (2) weak attributes. Strong attributes can be utilized to determine the occupation directly, i.e., uniforms, hats. Weak attributes, i.e., skin color and fraction, gender, age are relevant factors to some occupations but not sufficient. For example, dark skin color and large bare skin area are critical cues for a marathon player. These nondeterminant attributes together play essential roles when uniforms or hats are not applicable to particular occupations. The skin feature extraction process is illustrated in Fig. 10.5.

In this section, support vector machine (SVM) [7] is adopted to learn the first four attributes listed in Table 10.1 by feeding it with four different features $\phi_1(x), \phi_2(x), \phi_3(x)$, and $\phi_4(x)$. HOG [9] and LBP [1] features with the dense grid scheme [19] and color histogram for each upper-body or hat poselet are extracted. We then concatenate all of them to form a long vector as $\phi_1(x)$ and $\phi_2(x)$, respectively. The skin color feature $\phi_3(x)$ is generated based on the color histogram of the detected skin area, and the skin fraction $\phi_4(x)$ is the fraction of bare skin area over the entire human body. Therefore, we can predict each attribute by $y = w_i^{\mathsf{T}} \phi_i(x) + b_i$, where y is the output of SVM, and w_i, b_i is the learned weights and bias for the feature of the ith attribute. For binary classification problem, y is either -1 or 1 as output label. Such "hard" labeling, however, may lead to information loss, if we see visual attributes in a probabilistic view. In fact, soft scores provide more smoothness than hard labeling, and have potential to capture subtle changes in attributes. Therefore in our system, we use probabilistic output of SVM [8] as our visual attribute value.

$\phi_5(x)$ and $\phi_6(x)$ devote to the last two visual attributes, i.e., gender and age. We first use "haar feature + Adaboost framework" to detect human face. Then both gender and age of each person are computed through the framework in [15]. For gender, we use single variable 0 or 1 to denote female or male while for age, we quantize it into four binary bins based on partition points [0 18 40 60 100]. However, there are the cases that faces are missed by the face detector due to arbitrary poses. Under this situation, we assign 0.5 to the gender attribute. Since most of collected photos are people in their young or middle age, people with faces missed are assigned to age bin of 18–40. Finally, all the probabilistic outputs of these SVM with input $\phi_1(x), \ldots, \phi_4(x)$, and gender and age attributes are concatenate as the attribute vector a for the later joint learning model.

10.4 Recognition with Context

10.4.1 Approach Overview

This section explains how to utilize contexts in a photo to recognize multiple people's occupations. Three kinds of contexts, namely, (1) spatial relations, (2) co-occurrence, and (3) background information, are introduced. In addition, we show that how people in a specific scenario are tightly connected by these contexts.

We start from the attribute vector a obtained from last section. A probabilistic discriminative model $p(o_i|a_i)$ could be built for each occupation category, where a_i is the ith person's attribute and o_i is the ith person's occupation. When a collection of photos of multiple people are under test, each person's occupation can be recognized independently. However, three contexts are inevitably missed. In our model, suppose there are N detected human bodies $1, \ldots, N$, we consider any two people i and j with occupations o_i and o_j in the same photo by spatial relation features f_{ij}, pairwise relation type $r_{o_{ij}}$, and background type b_{ij} (Their detailed meanings are explained in later sections), corresponding to three contexts mentioned above.

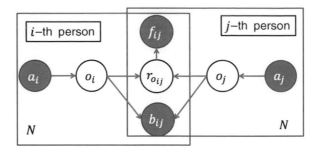

Fig. 10.6 Graphical model of the joint occupation recognition framework

Table 10.2 Notations of the graphical model

a_i: ith person attributes	A: all persons attributes
o_i: ith person occupation	O all occupations
$r_{o_{ij}}$: relation type between two occupations	R: all relation types
f_{ij}: spatial relation features between two people	F: all spatial relation features
b_{ij}: background features shared by two occupations	B: all background features
$m_i = j$: jth occupation is assigned to ith person	M: hidden variable for assignments

Intuitively, two people's occupation categories determine their pairwise relation type $r_{o_{ij}}$, and $r_{o_{ij}}$ therefore determines the spatial relation features f_{ij}. In addition, two people's occupation categories determine the background type b_{ij}. Take the occupation "waiter" for example. The pairwise relation type of a waiter and a customer is "service provider-customer[1]," and this infers that the customer should sit and the waiter should stand by in a short distance. Their occupations jointly decide the background information—indoor environment. Based on the analysis, our problem is finally formulated as a graphical model in Fig. 10.6 and its relevant notations are shown in Table 10.2. Our aim is to maximize the following conditional probability $p(O, R|A, F, B)$, which can be written as:

$$\frac{p(O, F, A, R, B)}{p(A, F, B)} \sim \sum_M p(O, F, A, R, B|M)p(M). \quad (10.1)$$

According to the graphical model, the former expression can be decomposed into the following formulation:

$$\sum_M \prod_{i=1}^N p(o_i|a_i) \prod_{i=1, j=1}^N p(f_{ij}|r_{o_{ij}})p(b_{ij}|o_i, o_j)p(r_{o_{ij}}|o_i, o_j)p(M), \quad (10.2)$$

where M is an indicator vector that explicitly models the occupation category assignment for each person. We use multiclass SVM and its probabilistic outcome

---

[1] We use "customer" as a latent occupation, though strictly it is not a real one.

[8] to model the probability $p(o_i|a_i)$ and use multinomial distribution to describe $p(b_{ij}|o_i, o_j)^2$. However, any other generative model or probabilistic discriminative model can be used here. Note that we use SIFT + BoWs model [13] to represent the background features, and b_{ij} is a 10×1 vector that bins the background feature through unsupervised learning, e.g., k-means. Since $p(r_{o_{ij}}|o_i, o_j)$ is always true in our designed model and hence equal to 1, we do not need to model this term in the learning or inference process.

Another key variable is $r_{o_{ij}}$ that denotes pairwise relation type between occupations o_i and o_j, which includes "service provider-customer" (e.g., waiter and customer), "customer-service provider" (e.g., patient and doctor), "colleague-colleague" (e.g., soldier and soldier), and "player-player" (e.g., marathoner and marathoner). We use three types of features to represent the spatial relation f_{ij} between two people, and for each relation feature we quantize them into several bins. Therefore, $p(f_{ij}^k|r_{o_{ij}})$ can be modeled as a multinomial distribution, where k indicates the kth feature. In addition, for simplicity, these features are assumed to be independent from each other, and therefore $p(f_{ij}|r_{o_{ij}})$ can be computed by the product of probability of each feature. The multinomial distribution parameters are smoothed by a Dirichlet prior. The relation features f_{ij} are detailed in the following:

Height Difference Height difference between two people reveals the specific social relation, especially when they are in some working environment. For example, customers sit beside a standing waiter. We use centers of torsos as the centers of people to calculate the height difference, and its absolute value. This value is normalized by the average width of human bodies in this photo. The height difference is quantized into five bins.

Distance Spatial distance between two people are related to their occupations. For example, students and teacher may be far away from each other in the classroom; doctors may stand close to each other when working. We use centers of torsos as the centers of people to calculate the distance. This value is normalized by the average width of human bodies in this photo. The distance is quantized into five bins.

Width Ratio Human body's width reveals the depth information to some extent if there are no babies included in the photo. This means the wider person locates close to the reader while the thinner one locates inside the photo. We use torso's width as person's width, and this ratio is quantized into five bins.

10.4.2 Learning

After tackle the decomposition for the proposed conditional probabilistic model, we need to learn the model parameters especially for $p(o_i|a_i)$, $p(f_{ij}|r_{o_{ij}})$, and $p(b_{ij}|o_i, o_j)$, where the first one is a linear model, and the second and the third ones are following multinomial distributions. Suppose we use a common parameter

---

[2] To simplify this part, we suppose b_{ij} only depends on o_i when we compute its probability.

θ to represent three sets of model parameters and our objective is to learn the parameter $\hat{\theta}$ through maximizing the following conditional probability:

$$\hat{\theta} = \arg\max_{\theta} p(O, R|A, F, B; \theta). \tag{10.3}$$

However, since a latent variable M is introduced in the likelihood function, we cannot immediately tackle the maximization problem in Eq. (10.1) via maximum likelihood. Instead, we use EM algorithm to solve it in an iterative way, which means we start our solution from a guess of model parameters θ. Like traditional usage of EM algorithms, e.g., mixture of Gaussian, in one iteration, we first calculate the probability of the assignment variable M given the current parameters θ^{old} in E step. We use uniform distributions as the guess of multinomial distributions, and $p(o_i|a_i)$ is initialized by ground truth labels and learned multiclass SVM model. Therefore, for a particular M^*, it can be updated by:

$$p(M^*|O, F, A, R, B; \theta^{\text{old}}) = \frac{p(O, F, A, R, B|M^*; \theta^{\text{old}})p(M^*; \theta^{\text{old}})}{\sum_M p(O, F, A, R, B|M; \theta^{\text{old}})p(M; \theta^{\text{old}})}, \tag{10.4}$$

where $p(O, F, A, R, B|M^*; \theta^{\text{old}})$ can be computed according to Eq. (10.2) based on current model parameters θ^{old}.

In the second part of one iteration, i.e., M step, we update the parameters in $p(o_i|a_i)$, $p(f_{ij}|r_{o_{ij}})$, and $p(b_{ij}|o_i, o_j)$ by maximizing the likelihood function in Eq. (10.1), which is achieved by combining Eqs. (10.2 and 10.4). Since the model parameters in these three probabilities are independent from each other, which means the derivative of one probabilistic model does not contain other type of parameters, we can update them one by one by computing one model parameter and fixing the other two.

10.4.3 Inference

After the model θ for the conditional probability in Eq. (10.1) is learned, it can infer the occupation categories for unknown test images. In the inference stage, we are given a test image containing n people. Then, we extract their attributes a_i, $i = 1, \ldots, n$, background type b_{ij}, and spatial relation features $f_{ij}, i, j = 1, \ldots, n$ and predict the occupation for each person. Since we use context information to constrain the occupation of each person in this image, the recognition results are no longer only dependent on attributes. Therefore, the occupation recognition based on this Bayesian network turns to be an inference problem as following:

$$M^* = \arg\max_M p(M|O, F, A, R, B), \tag{10.5}$$

where M is still the assignment vector, but with length of n.

Table 10.3 Average precision (%) for attributes of uniforms

Uniform	Chef	Clergy	Construction labor	Doctor	Firefighter	
Average precision	64.8	48.7	57.6	42.2	45.6	
Uniform	Lawyer	Mailman	Policeman	Soldier	Waiter	None
Average precision	65.1	29.9	42.9	59.2	32.8	18.7

10.5 Experimental Results

In this section, the proposed framework is evaluated step by step. First, we evaluate the effectiveness of the proposed visual attributes that represent semantic features. Second, we demonstrate that the proposed joint learning framework works better than the state-of-the-art method in [29]. Note that we use different partitions of the data for the experiments of visual attributes and joint learning framework, respectively. For visual attributes, images from each occupation category with one person in each photo are used for training and test while for the joint learning framework, only images with more than one person in each photo from each category are used for training and test. Note that in the joint learning framework, we do not need to label for photos with multiple people since the model parameters are learned in a fully automatic way.

10.5.1 Evaluation of Designed Attributes

For each attribute, half of images with this attribute are used as training and the other half as test. The negative samples of each attribute are selected in two different ways. First, for strong attributes, i.e., uniforms and hats, we use different uniforms or hats as negative samples. For example, if policeman uniform is the positive sample, then all uniforms except policeman's are negative samples. Similarly, half of them for are used for training and the other half for test. Second, for weak attributes, negative samples can be any people without these attributes. We define positive sample of skin color and fraction as white skin and long sleeves clothing and pants. In addition, we define male and people in a specific age range as positive samples. Since there are no gender and age labels available for our collected data, we manually label them for the test. We use one-to-rest binary classification strategy to test each attribute. The probabilistic outputs from SVM are used to compute precision and recall. We finally obtain the average precision for each attribute shown in Tables 10.3, 10.4, and 10.5, including strong and weak attributes.

From Tables 10.3, 10.4, and 10.5, it is observed that hats and uniforms of some occupations are very discriminative, e.g., soldier, doctor, and chef. However, some of them are not, e.g., mailman, construction labor. Moreover, we can see that weak attributes achieve relatively high performance, though they cannot directly determine

Table 10.4 Average precision (%) for attributes of hats

Hat	Doctor	Soldier	Chef	Policeman	Fire fighter	Construction labor	Without hat
Average precision	47.5	65.8	95.4	67.3	46.0	31.5	66.0

Table 10.5 Average precision (%) for weak attributes

Weak attributes	Skin color	Skin fraction	Gender	Age (0–18)	Age (18–40)	Age (40–60)
Average precision	78.1	65.4	66.9	71.3	63.5	42.1

Table 10.6 Experimental results of average precision (%)

	Chef	Clergy	Construction Labor	Customer	Doctor
Background features	10.3	10.8	11.4	7.4	9.6
Method in [29]	40.8	34.2	42.8	19.2	44.9
Person-level recognition	40.6	34.6	43.7	21.3	45.2
Joint recognition	42.3	35.1	46.1	27.6	48.9
	Firefighter	*Lawyer*	*Mailman*	*Marathoner*	*Policeman*
Background features	8.3	31.7	19.7	12.8	9.1
Method in [29]	31.3	59.1	21.8	48.2	18.4
Person-level recognition	30.3	57.6	23.1	52.1	20.1
Joint recognition	36.7	60.1	27.4	57.3	21.5
	Soccer player	*Soldier*	*Student*	*Teacher*	*Waiter*
Background features	28.8	31.5	14.8	7.8	17.6
Method in [29]	48.2	60.1	21.5	13.6	20.6
Person-level recognition	57.1	68.9	20.1	13.0	21.5
Joint recognition	60.2	74.7	25.0	15.2	28.6

The **average performance** of these methods are: Background features (15.4 %), method in [29] (35.0 %), person-level recognition (36.6 %), joint recognition (40.4 %)

the occupation category. Therefore, it is reasonable to use both high-performance but low-relevant and low-performance but high-relevant attributes to construct the person-level occupation classifier.

10.5.2 Evaluation of Joint Learning Framework

In our joint learning framework, the person-level occupation classifier, spatial relations, co-occurrence, and background are jointly considered through the Bayesian network discussed in the last section. In Table 10.6, comparisons with three other highly relevant methods are shown. First, in background features, we use SIFT + BoWs model to train a multiclass SVM for all occupation categories, and use background features in test images as inputs to determine people's occupations in the image. Second, the person-level recognition simulates the method in [29], but

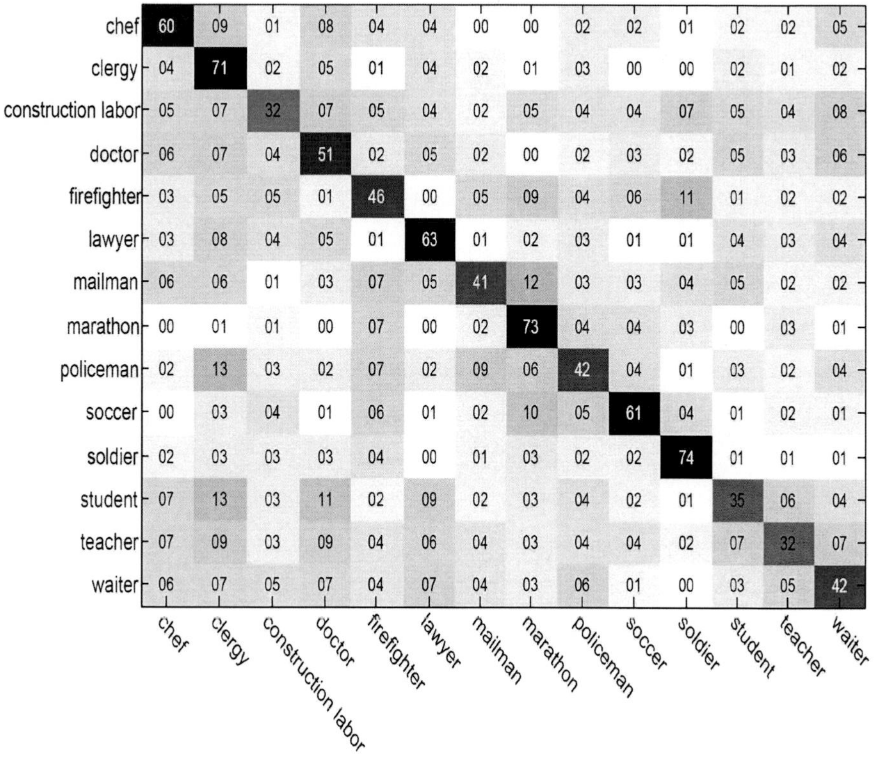

	chef	clergy	construction labor	doctor	firefighter	lawyer	mailman	marathon	policeman	soccer	soldier	student	teacher	waiter
chef	60	09	01	08	04	04	00	00	02	02	01	02	02	05
clergy	04	71	02	05	01	04	02	01	03	00	00	02	01	02
construction labor	05	07	32	07	05	04	02	05	04	04	07	05	04	08
doctor	06	07	04	51	02	05	02	00	02	03	02	05	03	06
firefighter	03	05	05	01	46	00	05	09	04	06	11	01	02	02
lawyer	03	08	04	05	01	63	01	02	03	01	01	04	03	04
mailman	06	06	01	03	07	05	41	12	03	03	04	05	02	02
marathon	00	01	01	00	07	00	02	73	04	04	03	00	03	01
policeman	02	13	03	02	07	02	09	06	42	04	01	03	02	04
soccer	00	03	04	01	06	01	02	10	05	61	04	01	02	01
soldier	02	03	03	03	04	00	01	03	02	02	74	01	01	01
student	07	13	03	11	02	09	02	03	04	02	01	35	06	04
teacher	07	09	03	09	04	06	04	03	04	04	02	07	32	07
waiter	06	07	05	07	04	07	04	03	06	01	00	03	05	42

Fig. 10.7 Confusion matrix of occupation recognition results (%)

using poselets + attributes instead of clothing to deal with pose variations. It shows that the proposed framework works comparably with the state-of-the-art method and sometimes even better. It becomes significant when people of this occupation tend to show arbitrary poses, e.g., sports player, soldier. Under this situation, clothing patch-based method is not stable and lots of background noise is filled into the clothing area. However, our poselet-based method still works well. On the other hand, method in [29] performs well when people of this occupation always show the nearly frontal pose, e.g., clergy, lawyer. In addition, the accuracy is enhanced by the interactive occupations, e.g., waiters and customers. We also find a significant improvement in occupations tending to show a group of people, e.g., soldier, marathoner, soccer player. This proves that our context-based Bayesian network is effective.

Although we see some improvements from the proposed model, in general, recognizing occupation from an image is not an easy task. There are many negative factors, to name a few: lack of significant attributes, e.g., students, teachers, diverse clothing features, e.g., mailman, week social context. Consequently, these categories are easily misclassified into other ones. To show the multiclass classification details over 14 categories, we also compute the confusion matrix by the recognition results and list it in Fig. 10.7. The number in ith row, jth column indicates the false alarm rate to

Fig. 10.8 Failed examples from the proposed method

jth class when recognizing ith class. From this table, we can summarize results, and hopefully reduce the error accordingly. For example, firefighters are easily misclassified as soldiers while students are randomly classified as other occupations since their attributes are not significant. Some misclassified samples are shown in Fig. 10.8.

10.6 Summary

In this chapter, a Bayesian network was proposed toward multiple people's occupation recognition in a photo. At low level, we adopted poselets to detect human body parts, therefore overcome the pose issue in real-world scenario. Second, we designed visual attributes that are favored by our problem to re-organize the relation between low-level features and high-level labels. Finally, social contexts and background information were integrated into the proposed probabilistic model to jointly infer multiple people's occupations in a photo. Evaluation results in this chapter demonstrated that our method is comparable to the state-of-the-art method, and performs better when social context can explicitly feature the group of people.

References

1. Ahonen, T., Hadid, A., Pietikäinen, M.: Face recognition with local binary patterns. In: European Conference on Computer Vision, pp. 469–481. Springer (2004)
2. Berg, T.L., Berg, E.C., Edwards, J., Maire, M., White, R., whye Teh, Y., Learned-miller, E., Forsyth, D.A.: Names and faces in the news. In: IEEE Conference on Computer Vision and Pattern Recognition, pp. 848–854. IEEE (2004)

3. Bourdev, L., Maji, S., Brox, T., Malik, J.: Detecting people using mutually consistent poselet activations. In: European Conference on Computer Vision, pp. 168–181. Springer (2010)
4. Bourdev, L., Maji, S., Malik, J.: Describing people: poselet-based attribute classification. In: International Conference on Computer Vision, pp. 1543–1550 (2011)
5. Bourdev, L., Malik, J.: Poselets: body part detectors trained using 3D human pose annotations. In: International Conference on Computer Vision, pp. 1365–1372 (2009)
6. Brox, T., Bourdev, L., Maji, S., Malik, J.: Object segmentation by alignment of poselet activations to image contours. In: IEEE Conference on Computer Vision and Pattern Recognition, pp. 2225–2232. IEEE (2011)
7. Burges, C.: A tutorial on support vector machines for pattern recognition. Data Min. Knowl. Disc. **2**(2), 121–167 (1998)
8. Chang, C.C., Lin, C.J.: LIBSVM: A library for support vector machines. ACM Trans. Intell. Syst. Technol. **2**, 1–27 (2011). Software available at http://www.csie.ntu.edu.tw/cjlin/libsvm
9. Dalal, N., Triggs, B.: Histograms of oriented gradients for human detection. In: IEEE Conference on Computer Vision and Pattern Recognition, vol. 1, pp. 886–893. IEEE (2005)
10. Desai, C., Ramanan, D., Fowlkes, C.: Discriminative models for multi-class object layout. In: International Conference on Computer Vision, pp. 229–236 (2009)
11. Farhadi, A., Endres, I., Hoiem, D., Forsyth, D.: Describing objects by their attributes. In: IEEE Conference on Computer Vision and Pattern Recognition, pp. 1778–1785. IEEE (2009)
12. Fathi, A., Hodgins, J.K., Rehg, J.M.: Social interactions: a first-person perspective. In: IEEE Conference on Computer Vision and Pattern Recognition, pp. 1226–1233. IEEE (2012)
13. Fei-Fei, L., Perona, P.: A bayesian hierarchical model for learning natural scene categories. In: IEEE Conference on Computer Vision and Pattern Recognition, pp. 524–531. IEEE (2005)
14. Fu, Y., Guo, G., Huang, T.: Age synthesis and estimation via faces: a survey. IEEE Trans. Pattern Anal. Mach. Intell. **32**(11), 1955–1976 (2010)
15. Gallagher, A., Chen, T.: Understanding images of groups of people. In: IEEE Conference on Computer Vision and Pattern Recognition, pp. 256–263. IEEE (2009)
16. Kankuekul, P., Kawewong, A., Tangruamsub, S., Hasegawa, O.: Online incremental attribute-based zero-shot learning. In: IEEE Conference on Computer Vision and Pattern Recognition, pp. 3657–3664. IEEE (2012)
17. Kumar, N., Berg, A., Belhumeur, P., Nayar, S.: Attribute and simile classifiers for face verification. In: International Conference on Computer Vision, pp. 365–372. IEEE (2009)
18. Lampert, C., Nickisch, H., Harmeling, S.: Learning to detect unseen object classes by between-class attribute transfer. In: IEEE Conference on Computer Vision and Pattern Recognition, pp. 951–958. IEEE (2009)
19. Lazebnik, S., Schmid, C., Ponce, J.: Beyond bags of features: spatial pyramid matching for recognizing natural scene categories. In: IEEE Conference on Computer Vision and Pattern Recognition, pp. 2169–2178. IEEE (2006)
20. Liu, S., Song, Z., Liu, G., Xu, C., Lu, H., Yan, S.: Street-to-shop: cross-scenario clothing retrieval via parts alignment and auxiliary set. In: IEEE Conference on Computer Vision and Pattern Recognition, pp. 3330–3337. IEEE (2012)
21. Maji, S., Bourdev, L., Malik, J.: Action recognition from a distributed representation of pose and appearance. In: IEEE Conference on Computer Vision and Pattern Recognition, pp. 3177–3184. IEEE (2011)
22. Naaman, M., Yeh, R., Garcia-Molina, H., Paepcke, A.: Leveraging context to resolve identity in photo albums. In: ACM/IEEE-CS Joint Conference on Digital Libraries, pp. 178–187 (2005)
23. Parikh, D., Grauman, K.: Relative attributes. In: International Conference on Computer Vision, pp. 503–510. IEEE (2011)
24. Patterson, G., Hays, J.: Sun attribute database: discovering, annotating, and recognizing scene attributes. In: IEEE Conference on Computer Vision and Pattern Recognition, pp. 2751–2758. IEEE (2012)
25. Ramanathan, V., Yao, B., Fei-Fei, L.: Social role discovery in human events. In: IEEE Conference on Computer Vision and Pattern Recognition, pp. 2475–2482. IEEE (2013)

26. Saleh, B., Farhadi, A., Elgammal, A.: Object-centric anomaly detection by attribute-based reasoning. In: IEEE Conference on Computer Vision and Pattern Recognition, pp. 787–794. IEEE (2013)
27. Shahbaz Khan, F., Anwer, R.M., van de Weijer, J., Bagdanov, A.D., Vanrell, M., Lopez, A.M.: Color attributes for object detection. In: IEEE Conference on Computer Vision and Pattern Recognition, pp. 3306–3313. IEEE (2012)
28. Shao, M., Li, L., Fu, Y.: Predicting professions through probabilistic model under social context. In: workshops at the 27th AAAI Conference on Artificial Intelligence (2013)
29. Song, Z., Wang, M., Hua, X., Yan, S.: Predicting occupation via human clothing and contexts. In: International Conference on Computer Vision, pp. 1084–1091 (2011)
30. Wang, G., Forsyth, D.: Joint learning of visual attributes, object classes and visual saliency. In: International Conference on Computer Vision, pp. 537–544. IEEE (2009)
31. Wang, G., Gallagher, A., Luo, J., Forsyth, D.: Seeing people in social context: recognizing people and social relationships. In: European Conference on Computer Vision, pp. 169–182. Springer (2010)
32. Wang, Y., Mori, G.: A discriminative latent model of object classes and attributes. In: European Conference on Computer Vision, pp. 155–168. Springer (2010)
33. Xia, S., Shao, M., Luo, J., Fu, Y.: Understanding kin relationships in a photo. IEEE Trans. Multimedia 14(4), 1046–1056 (2012)
34. Yuan, X., Yan, S.: Visual classification with multi-task joint sparse representation. In: IEEE Conference on Computer Vision and Pattern Recognition, pp. 3493–3500. IEEE (2010)
35. Zhao, W., Chellappa, R., Phillips, P., Rosenfeld, A.: Face recognition: a literature survey. ACM Comput. Surv. 35(4), 399–458 (2003)

Index

A
Action recognition, 75, 78, 95, 96
Aging pattern, 150–158, 167, 171, 172
Aging pattern subspace, 157, 158
Attribute recognition, 136

C
Classification, 149, 152, 162
Clustering, 89, 92
Community detection, 44–46, 52, 57–59, 65, 66
Community profiling, 60, 61
Conditional random field, 76, 92
Conditional topic random fields, 82
Cross-domain, 6, 9, 18

D
Dense subgraph mining, 57
Discrimination, 105
Discriminative decision trees, 100, 103, 112

F
Face recognition, 117–119, 127
Facial beauty attribute, 136
Facial image editing, 133, 135
Fine-grained categorization, 95, 98
Foursquare, 43, 45–50, 52, 57, 58, 63, 65

G
Group photo labelling, 175

H
Heterogeneous hypergraph, 52, 53
Human identification, 175
Human tracks (or) tracklets, 79, 81, 87, 92
Human–object interaction, 95, 97, 108

K
Kin relationship, 175–178, 181, 183
Kinship similarity, 175, 177–181, 183, 185, 186, 188, 189

L
Label distribution, 151, 159–162, 166, 171, 172
Label distribution learning, 151, 161, 162, 171
Location-based social networks, 43

M
Maximum entropy model, 164, 171

O
Object bank, 110
Object interaction feature, 81
Occupation recognition, 191, 193, 200, 204

P
Pairwise interaction features, 82, 92
PCA, 154, 155
Popularity, 3–6, 8, 13–16, 19
Proxemics, 82

Y. Fu (ed.), *Human-Centered Social Media Analytics*,
DOI: 10.1007/978-3-319-05491-9, © Springer International Publishing Switzerland 2014